THE MAL-OBSERVATION REPORT

Richard Hodgson
S. J. Davey

New York

DR. RICHARD HODGSON
This "spirit photograph" was taken to show that a face may be made to appear over jewelry (in this case, the watch chain) as the result of fraudulent manipulation of the plates. (Hereward Carrington's *The Physical Phenomena of Spiritualism*, 1920.)

THE MAL-OBSERVATION REPORT

THE POSSIBILITIES OF MAL-OBSERVATION AND LAPSE OF MEMORY FROM A PRACTICAL POINT OF VIEW

— and —

MR. DAVEY'S IMITATIONS BY CONJURING OF PHENOMENA SOMETIMES ATTRIBUTED TO SPIRIT AGENCY

Foreword by
Dr. Matthew L. Tompkins,
BA, MSc, DPhil, MMC

Cur*i*ous
PUBLICATIONS
New York

Published by Curious Publications
697 Third Ave. #358
New York, NY 10017
curiouspublications.com

Copyright © 2026

Library of Congress Control Number: 2025923560

ISBN-13: 979-8-9914395-8-9

This book faithfully reproduces the original 1887 and 1892 papers by Richard Hodgson and Samuel J. Davey, as published in *Proceedings of the Society for Psychical Research*. The text and images are in the public domain.

Cover image: *'Twixt Two Worlds: A narrative of the life and work of William Eglinton*, by John Stephen Farmer, The Psychological Press, 1886.

Inside front and back cover images: *Psychography: Marvelous Manifestations of Psychic Power Given though the Mediumship of Fred P. Evans*, by James J. Owen, Hicks-Judd Co., 1893.

Printed and bound in the United States of America.

CONTENTS

Foreword	7
The Possibilities of Mal-Observation and Lapse of Memory from a Practical Point of View	51
Mr. Davey's Imitations by Conjuring of Phenomena Sometimes Attributed to Spirit Agency	207
Further Reading and Additional Resources	283

Foreword
by Dr. Matthew L. Tompkins,
BA, MSc, DPhil, MMC.

If trance, the involuntary life, and human testimony, were understood universally as they are now beginning to be understood by students of the nervous system, there would not, could not be a spiritist on our planet; for all would know that spirits only dwell in the cerebral cells—that not our houses but our brains are haunted. (Beard, 1879, p. 67)

SPEAKING as someone with experience being a fake psychic, I would like to invite you to think of this book as an act of mediumship—we're conveying the words of the dead and conjuring a largely forgotten study of human memory. After all, what is a book if not a device for transmitting thoughts directly into your mind? This text is a reproduction of two scientific papers that were originally published in 1887 and 1891: "The possibilities of mal-observation and lapse of memory from a practical point of view" (which I'll refer to subsequently as "The Mal-Observation Report") and "Mr. Davey's Imitations by Conjuring of Phenomena Sometimes Attributed to Spirit Agency." The experimenters, Richard Hodgson and Samuel John Davey, conducted a series of fake séances, using magic trick methods. They invited 'sitters' to observe these séances. The sitters were not charged any money, but rather the price of admission was that they were asked to write a letter describing the events that they had experienced in as much detail as they could remember. Because the experimenters choreographed the séances themselves, they had a detailed understanding of the real events that occurred, which they could compare to the collected witness statements. Their results were dramatic: Witnesses consistently made systematic errors in their reports—failing to notice or remember key details, and even reporting events that could not possibly have occurred. Many readers, including eminent scientists at the time, were highly skeptical of the results. Some critics of the report even argued that the experimenters themselves were lying about the study, and that they really did possess genuine psychic abilities. The idea that the experimenters were communing with the spirits of the dead was considered more plausible than the idea that sober, healthy, honest witnesses could commit such dramatic errors. As a result, Hodgson felt compelled to publish a follow-up paper re-asserting that the séances were genuinely faked. To bolster his argument, he revealed more details of how they accomplished their tricks using conjuring methods rather than supernormal powers. In the intervening 138 years, many of the report's conclusions about the fallibility of human memory have been borne out in more formal academic studies—we now have established terms like inattentional blindness and reconstructive memory to describe some of the psychological phenomena revealed by the experiments. The scientific community has not come to embrace the claim that spiritual mediums can commune with the dead, but scientists have come to accept that honest observers can make surprising errors in how they see and remember events. Looking back, the Mal-Observation Report is truly groundbreaking work, but, ironically, it has been largely forgotten and unnoticed by contemporary researchers and writers. You are extremely unlikely to hear about this study in a university psychology lecture, or to see it referenced in conventional textbooks on memory, perception, or eyewitness testimony. But, nonetheless, as a magician and experimental psychologist, I'd like to make the case that this work is well worth remembering.

Within this book, we have reproduced the original text of the Mal-Observation Report, including the author's original theoretical justifications and conclusions and all the published witness statements. We've also included the text of the follow-up paper, which sought to assuage critics by providing additional details about the methods used to create the illusions described. Together the texts form a kind of ghost story. But this story is not about haunted houses, it's about haunted brains.

To help set the stage, I'll be taking the next few pages to contextualize the words that we have reproduced—both in terms of when they were written and also their continuing relevance today.

An argument from authority

First, a bit about my own qualifications: I am both a performing magician and an experimental psychologist. My interest in magic began in the traditional fashion, as a small child I had a formative experience watching a magician perform seemingly impossible feats with a handful of silver dollars. In his hands, the coins appeared to defy the laws of physics: winking in and out of existence and inexplicably floating in midair. Shortly afterwards, my parents gifted me with a copy of J.B. Bobo's *Modern Coin Magic* (1952). This book served as my gateway into the weird world of performance magic methods. Before long, my very first job was performing as a magician at other children's birthday parties. In hindsight, this was a particularly brutal crash-course in applied developmental psychology. I do not perform professionally for children anymore, but my brief time in the business instilled me with a deep lasting respect for children's entertainers. As I grew older, I moved on to perform close-up strolling magic for adult audiences.

When I eventually attended university to study psychology as an undergraduate, I continued working semi-professionally as a magician. I studied at the State University of New York at Geneseo, coincidentally, not far from the historical birthplace of Modern Spiritualism in Hydesville, New York. In my courses and my lab work, I was increasingly struck by parallels between magic methods and the methods of psychological experiments designed to explore human perception, memory and reasoning. I became aware of a small but vibrant literature on the psychology of magic spearheaded by researchers like Gustav Kuhn, Richard Wiseman, and Peter Lamont. As I progressed into my graduate studies in experimental psychology, I kept performing, in parallel, as a magician, and I steadily worked to bring those two aspects of my life closer together. When applying for my doctoral position at Oxford, I performed some close-up coin magic for the interview committee to illustrate my proposed research (Ingraham, 2016). My doctoral thesis focused on the historical and contemporary ways that magic methods can be adapted to advance psychological research. In exploring the history of using magic to study psychology, I learned that, while the current "renaissance" of science of magic research was relatively new, the practice could be traced back to the earliest days of experimental psychology when nascent psychologists attempted to use the idea of illusions to explain spiritualistic claims. The Mal-Observation Report is the earliest record I have been able to find of researchers employing magic tricks to help them conduct behavioral research.

Of particular relevance to this current text, my first gig after completing my doctorate was acting as a (fake) spiritual test medium. I was hired by Gustav Kuhn, then himself a Reader in Psychology at Goldsmiths University, to help administer a rather peculiar classroom exercise for his research students: Gustav would present me to his students as an individual who might potentially have real psychic powers. The students' job was to administer tests to determine if my powers were genuine. My job was to cheat as horrifyingly as possible on the tests without getting caught. At the end of the exercise, the students were asked to assess the validity of my supposed psychic gifts, and the majority consistently asserted that their tests had empirically proven that my powers were real. Generally, as an educator and science communicator, I try to be very careful about how I discuss and present pseudoscientific claims and techniques. The main reason I felt comfortable with this arrangement was because, at the end of every exercise, Gustav and I would explicitly reveal that I was in fact a magician, who had been cheating. We would then explain that the purpose of the exercise was to demonstrate the methodological challenges of dealing with deception on the part of research participants. During one of these sessions, a reporter named David Baker, from Wired magazine, sat in on the class. Afterwards, he conducted interviews with the students. To our chagrin, we learned from his reporting that, for some of the students, our fake demonstrations were actually more convincing than our real explanations about how I was cheating. One of the students was quoted in the article as saying, after the explanation, "I still think there was something weird about that guy. He could read minds" (quoted in Baker, 2019, p. 135). Not my proudest moment as an educator... I have continued to use magic and deception as tools in my research, although I'm now even more careful about dehoaxing participants at the end of experiments.

These days, I work as a researcher in the Choice Blindness Lab at

Lund University in Sweden. My main role is to design and implement fake mind control machines. Seriously. We use these fake machines to study how people perceive and misperceive technology (Olson et al. 2023; Tompkins, 2023). In our experiments, I deliberately employ some of the exact same techniques of fraudulent mediumship that S.J. Davey used back in the 1880s. The main difference between my tricks and Davey's is that, instead of pretending to be a medium for fictional spirits, I pretend to be a medium for fictional artificial intelligences. When we frame extraordinary experiences, like thought reading, divination, and mind control, as originating from advanced technologies, the vast majority of our participants express confidence that our non-existent AI systems are real, just like the participants in the original Mal-Observation study more than 100 years ago. These results have been particularly professionally satisfying for me. In my work, I often speak about the history of spiritualism and mediumistic fraud, and I find that modern audiences are often inclined to view such fraud as something of a historical anachronism. They seem to think that such deceptions were only effective because they were perpetrated on less educated people from the unenlightened past. But of course, this was also what the Victorians thought. I find it gratifying to generate fresh empirical evidence that the same magic trick methods that fooled Hodgson and Davey's participants back in the 1880s remain highly deceptive to modern audiences. In summation, I have spent my life and career studying and practicing illusions and deceptions, and that's why you should trust me.

Setting the Scene:
The Historical Context of the Mal-Observation Report

The Mal-Observation Report was produced during the height of what is now known as Modern Spiritualism. This socio-cultural religious movement differed from more traditional religious practices in that it professed to offer empirical miracles. Rather than inviting believers to take miraculous stories on faith, spiritualists offered to be able to demonstrate phenomena like the direct communication with the souls of the dead. Modern Spiritualism arguably has roots in mystical practices dating back centuries, and it can be directly linked to prophetic writings of authors like Emmanuel Swedenborg (1758) and Andrew

Jackson Davis (1847). But there is a particularly convenient narrative origin point: An event that was documented at a particular time and place: The 31st of March 1848 (a little after 8pm) in the family home of John Fox, located in the town of Hydesville in upstate New York. It was there and then that John's young daughters, Kate (age 11) and Maggie (age 14), allegedly made contact with a spiritual entity that manifested itself through knocking noises or "raps," that seemed to have no earthly origin. The family and their neighbors were unable to identify any visible natural source of the raps. And the girls developed a primitive code that allowed them to communicate with the entity. It eventually identified itself as the spirit of a peddler who had been murdered and buried in their basement (Cadwallander, 1917; Lewis, 1848; Weisberg, 2004, see Nickell, 2008 for a more skeptical reinterpretation of the events). News spread rapidly of the girls' miraculous discovery, and the next year, on November 14, 1849, they displayed their mediumistic skills to a packed (paying) crowd at Corinthian Hall in the nearby city of Rochester. Proponents of spiritualism have argued that the Fox Sisters' development as mediums represented the dawning of a new age, characterized by a thinning of the veil between our world and the world of spirits. Skeptics have suggested that the popularity of the sisters' act inspired other fraudulent mediums to take financial advantage of a gullible public. Either way, the practice of spiritualism and mediumship would spread rapidly around the United States and the world. The movement maintains believers to this day; I personally used to regularly attend séances as an observer at spiritualist churches throughout London while I was writing my doctoral thesis. But on the topic of skepticism, the Fox Sisters faced accusations of trickery and fraud from the beginning. And Maggie eventually confessed publicly that the original raps were actually produced by an apple tied to a string that the girls would knock against their bedframe (Davenport, 1888). Skeptics felt vindicated, but believers were skeptical of the confession, and spiritualist beliefs and practices continued to spread.

The empirical claims of Modern Spiritualism drew the attention of many academics, scientists, and intellectuals. In 1882, the Society for Psychical Research (SPR) was formed in Cambridge, UK with the express purpose of scientifically assessing spiritualist claims. Their cases involved phenomena like mediumistic séances, telepathy, haunted houses, apparitional experiences, and the idea of the survival of the soul after bodily death. If any such spiritualist claims could be verified as real, then they would represent a genuine scientific revolution. If they could be shown to be false, then they could provide fascinating insights into the psychology of fraud, illusion, deception and self-deception. Spiritualist practices had rapidly evolved beyond the initial rapping noises demonstrated by the Fox Sisters; mediums developed a variety of alternative methods of communication. In some instances, séances might involve the physical manifestations of spirits, in the form of disembodied limbs or entire figures. But the communication style that would become the subject of the SPR's Mal-Observation Report was the practice of slate writing, or psychography (Oxon, 1882). In slate writing séances, the medium would claim to receive written communications from spirits that would manifest inexplicably on small handheld chalkboards. While out of fashion now, such slates were common everyday objects at the time, particularly in school classrooms. You can think of them as a sort of Victorian iPad. Rather than consuming paper for their daily assignments, school children would conduct written work on their re-usable slates. At a typical slate writing séance, the medium would invite sitters to examine one or more slates (potentially provided by the sitters themselves), and then, under seemingly impossible conditions, written messages or drawings would appear to manifest upon the previously blank slates. In the words of one believer:

> *What do we mean by "independent slate-writing"? I understand that term to signify the formation of legible letters and words on a slate by a pencil which no one touches while the writing is being done.* (Cous, 1892, p. 628)[1]

Perhaps the most notorious slate-writing medium in history is Henry Slade (Brown, 2017), whose globe-trotting career is intimately

---

[1] In his assertively titled article, "Independent Slate Writing: A Fact of Nature," Cous was seeking to draw a contrast with the phenomena of automatic writing, which was a distinct phenomenon that occurred when a medium openly held a pencil and wrote, but would assert that the words produced were generated unconsciously, potentially attributable to external spiritual forces that had taken direct control of their body.

linked with spiritualism, scepticism, and the emergence of experimental psychology as an academic discipline. Slade got his start working as a medium in America. In the 1870s, he embarked on a world tour with the purpose of demonstrating the reality of his spiritual talents to both the public and men of science. However, his journey got off to a rocky start. When he arrived in London, he was promptly arrested and tried for witchcraft. Technically, he was not being charged with *practicing* witchcraft, but rather he was accused of defrauding the British public by duping people into believing he had supernatural powers. His sensational trial drew international media attention (e.g. Milner, 1996). Slade was ultimately convicted, but then that conviction was overturned on a legal technicality. Before he could be re-tried, Slade fled to continental Europe, and eventually ended up in Leipzig, Germany where he caught the attention of Johann Zöllner, an astrophysicist with a strong interest in spiritualism. Zöllner conducted a long series of experiments with Slade and documented a variety of extraordinary phenomena that occurred in the presence of the medium: Not only independent slate writing, but magnetic anomalies, the production of impossible knots, the materialization and dematerialization of objects, and even the manifestation of otherworldly spirits. Zöllner was particularly delighted to report that he had the opportunity to shake hands with a ghostly disembodied limb that he described as "a friend from another world" (Zöllner/Massey, 1880). Zöllner published his results and conclusions, declaring that he had developed a new branch of science: "Transcendental Physics." His theoretical framework was designed to explain Slade's mediumship in terms of 4th-dimensional geometry (Zöllner/Massey, 1880). At one point, Zöllner invited his junior colleague, Wilhelm Wundt, to sit-in on one of Slade's demonstrations. Controversy ensued when Wundt published a brief open letter proposing his own competing theory to explain Slade's power: Perhaps Slade was simply deceiving Zöllner with conjuring tricks (Wundt, 1879)? Zöllner was outraged. He publicly called for Wundt to be arrested for slander, and even hypothesized that perhaps Slade's magnetic powers had disrupted Wundt's brain. Ultimately, Slade continued with his controversial tour. Wundt was not imprisoned, and he actually went on to establish the world's first ever officially recognized psychology laboratory (Marshall & Wendt, 1980, Tompkins, 2017). Zöllner's

Transcendental Physics has not been widely adopted, but Wundt is recognized today as one of the founding fathers of experimental psychology. His open letter about Zöllner's experiments with Slade, "Spiritualism as a Scientific Question," raised an interesting idea that psychologists might turn to magic methods to help provide naturalistic explanations for accounts of extraordinary, supernatural-seeming, occurrences. Wundt himself never seriously engaged with this idea beyond vaguely suggesting that Slade might have been guilty of employing covert "jugglery" (i.e. sleight of hand). But other researchers, like Hodgson and Davey, and later even myself and my contemporary colleagues, would go on to directly explore the psychological relations between magic tricks, illusions, experiences, and belief.

By 1887, partially as a result of Zöllner's scientific publications about Slade, the SPR was particularly interested in the idea of independent slate writing. By that time, a sort-of successor to Slade had emerged in Britain, a medium named William Eglinton. Witnesses at Eglinton's séances described how, like Slade, he could mysteriously produce writing on borrowed slates that provided answers to questions he himself had no earthly way of knowing. Many spiritualists considered slate writing phenomena to be among the most definitive empirical proofs of spiritualism. A prominent member of the SPR, Frank Podmore, wrote of Eglinton's alleged abilities: "In the history of the movement, no physical manifestation had ever won such universal recognition" (Podmore, 1902, p. 205). The SPR compiled numerous accounts of Eglinton's séances (e.g. Hodgson, 1886). Skeptics, including a number of professional magicians, argued that the same results could be produced by using entirely naturalistic magic trick methods. Eglinton's mediumship generated a fierce debate within the society. One of his most outspoken critics was Eleanor Mildred Sidgwick, a Cambridge academic and an another influential member of the SPR. She proposed that the SPR should never seriously consider the powers of mediums who had ever been credibly accused of fraud (Sidgwick, 1886a; 1886b). And she further stated that this rule ought to be applied to Eglinton. She noted how Eglinton had allegedly been caught with physical evidence that he was faking spiritual phenomena. Specifically, he had been found to be carrying around a collection of suspicious props, including drapes and fake beard, that physically bore an un-

WILLIAM EGLINTON
English medium seen with his closed slate on the table.
(*'Twixt Two Worlds: A Narrative of the Life and Work of William Eglinton*, 1886.)

canny resemblance to Abdulah, a spirit who would regularly feature as a physical manifest during Eglinton's mediumistic demonstrations. Sidgwick went on to suggest that the extraordinary accounts of Eglinton's phenomena should be dismissed by serious researchers as resulting from a combination of conjuring-style deceptions and mal-observations on the part of witnesses. In other words, she proposed that, despite the amazing stories of his supernatural powers, Eglinton used mundane magic trick methods that were both misperceived and/or misremembered by his audience. Eglington and his supporters in the SPR were upset by these accusations. Charles Carleton Massey, a barrister and devout believer in spiritualism, published a scathing reply, in which he argued that Sidgwick's arguments about mal-observation were absurd (Massey, 1886). Massey argued that no reasonable person could ever mistake conjuring tricks for genuine psychical phenomena, and that no intelligent honest witness could commit errors of testimony due to lapses in observation or memory that could possibly account for all the wondrous stories of Eglinton's mediumship. In response to Massey's counter-argument, Eleanor Sidewick sought out the assistance of Angelo John Lewis, a barrister who also worked as a professional magician. He performed and wrote under the stage name Professor Hoffman. They reasoned that if Eglinton was resorting to conjuring tricks, a magician would be much more capable of penetrating his illusions than someone who was untrained in conjuring methods. Not only did Lewis take up the task of analyzing many written accounts of Eglinton's seances, but he also personally met with Eglinton in hopes of experiencing the phenomena for himself. In his analysis of the reports, Lewis wrote that witnesses of magic tricks cannot necessarily be trusted to accurately describe events as they occurred, but rather describe the events that they imagined had happened. In fact, a good magic performance is often scripted and choreographed to maximize the likelihood of this happening. Lewis concluded that, based on his experiences as a magician, many first accounts of Eglinton's phenomena seemed to suggest that the witnesses might be unwittingly committing systematic errors of testimony by misperceiving or misremembering key elements of Eglinton's performances. But it was impossible to prove anything definitively from the testimonies. He and Sidgwick hoped that his personal visits to Eglinton might result

The Mal-Observation Report

How the writing is done. (*Psychography: Marvelous Manifestations of Psychic Power Given though the Mediumship of Fred P. Evans*, 1893.)

in more concrete proof that the medium was resorting to conjuring tricks. After all, Lewis was much more qualified to identify conjuring tricks than the average witness. Unfortunately for them, this is where their investigation stalled. Lewis visited Eglinton twelve times, but each time "the spirits obstinately declined to manifest." Such outcomes were not unknown in attempted demonstrations of mediumship; they even had their own term: a blank seance. These resulted in something of an epistemological impasse in the rhetorical struggle between Eglinton's sceptical critics and his proponents. On one hand, the sceptics could argue that these blank seances were evidence that Eglinton was attempting to conceal his conjuring methods from Lewis. They could point out that Eglinton's 'powers' worked quite consistently until he was confronted by someone who was likely to catch him in his deceptions. If Eglinton had been resorting to trickery, of course it would make sense for him to avoid risking exposure when he knew he was being observed by a well-known and respected professional conjurer. But on the other hand, Eglinton's proponents could likewise feel vindicated by the blank seances. After all, if Eglinton was a fraud who used magic tricks, then he ought to be able to manifest his fake powers at any time. The fact that he sometimes entirely failed to produce phenomena was seen as evidence that his powers were genuine, because they were beyond his conscious control. Hodgson and Davey's report represented a novel methodological solution to this problem.

Hodgson and Davey's Mal-Observation Report

Hodgson and Davey's Mal-Observation Report describes a scientific investigation that was developed as a way around the empirical problem of investigating the reliability of testimonies related to allegedly genuine spiritualist mediumship. By using magic trick methods to simulate slate writing phenomena, they avoided the problem of blank séances that had plagued Lewis' attempts to observe Eglinton. Furthermore, by relying exclusively on magic trick methods performed by an openly fake medium who made no claim to supernormal abilities, they hoped to avoid the potential criticism that genuine psychical phenomena might have occurred during their experiments. Hodgson took specific care to declare to readers of the report that even though the participants' testimonies might seem "marvelous enough" to sup-

An Apparition Formed in Full View.
(*'Twixt Two Worlds: A Narrative of the Life and Work of William Eglinton*, 1886.)

port the hypothesis of occult agency, "the 'psychographic' phenomena described in the following records are conjuring, and only conjuring, performances" (p. 400). "The object of the notes," he went on to write, was not to educate readers in any particular conjuring methods, but rather "to show to investigators the kind and degree of mistakes which may be made by educated and intelligent witnesses in recording their impression of a performance the main lines of which are planned with the deliberate intention of deceiving them, but few, if any, of the details of which can be described as absolutely fixed" (p. 402).

The Mal-Observation Report represents the synthesis of Davey's conjuring expertise and Hodgson's first-hand experience debunking professional mediums. Richard Hodgson was born in Melbourne, Australia and obtained a degree in Law at the University of Cambridge (Baird, 1949). During his time at Cambridge, Hodgson befriended Prof. Henry Sidgwick and joined the recently formed SPR. As a member of the society, Hodgson developed a reputation as a talented detective and debunker of spiritualist mediums. He spearheaded one of the SPR's first major investigations into spiritualism when he travelled to India to investigate Countess Helena Petrova de Blavatsky, a Russian occultist who had attracted a significant following based on her alleged supernatural abilities. She claimed that she was channeling the spiritual teachings of dead mystics. She also claimed that she could project her consiousness outside of her body in the physical form of a spirit. Hodgson spent two months in India attempting to verify Blavatsky's claims. In his eventual report, which he published through the SPR in 1885, he declared that the phenomenon that she claimed credit for were entirely fraudulent. The letters she claimed to receive from spirits of the dead were written in her own hand, and the spirit forms she took were simply her human accomplices wearing disguises. He argued that her followers were "excessively credulous and deficient in observation," and that she was actually the center of a "huge fraudulent system" (Hodgson, 1885). Hodgson's experiences investigating Blavatsky arguably helped spark his initial interest in the relationship between magic tricks and eye-witness testimony, and after meeting S.J. Davey, he would go on to co-develop a new experimental paradigm to empirically explore these ideas.

Hodgson's co-experimenter, Samuel John Davey, was an amateur

SAMUEL J. DAVEY
(*A Guide to the Collector of Historical Documents, Literary Manuscripts, and Autograph Letters, etc.*, 1891. This image was rediscovered by Richard Wiseman in 2025.)

magician. Before he began practicing magic, Davey had been a true believer in the reality of spiritualist phenomenon. He wrote that he initially came to his beliefs after seeing a recently deceased friend in a dream, and that his beliefs were further inspired by reading an English translation of Johann Zöllner's *Transcendental Physics* (1881). He also personally witnessed the apparently miraculous talents of the medium William Eglinton. Davey was particularly fascinated by the phenomenon of slate writing. In the introduction to the Mal-Observation Report, he confessed that, for a time, he actually came to believe that he himself possessed genuine supernatural talents. He experimented by trying to receive messages from spirits by leaving blank slates in locked drawers in his house. He was astonished upon examining some of the slates to find that messages had appeared on the slates that he himself had cleaned and locked away. Eventually, he came to realize that his friends were hoaxing him by secretly marking the slates themselves through entirely natural means. This experience of being tricked led Davey to begin studying conjuring. As a consequence, he came to appreciate that many of the effects that he had previously considered to be miraculous were actually attributable to magic trick methods. Davey soon progressed beyond academic curiosity about magic and began putting his conjuring knowledge into practice. He approached the Society for Psychical Research to demonstrate his developing conjuring skills. And after witnessing Davey's tricks, Richard Hodgson and Eleanor Sidgwick realized that Davey's fake mediumship represented an opportunity to develop a new method to empirically investigate testimonies of alleged psychic phenomena.

The Mal-Observation Report opens with a general introduction written by Hodgson describing his own experiences investigating psychic fraud. He professed having once been very confident that eye-witness testimonies of séances represented examples of genuine phenomena. But, over time, as he learned more about the possibilities of trickery and conjuring methods, he grew more and more skeptical. Hodgson directly addresses both Alfred Russell Wallace (1896) and Charles Carleton Massey's (1886) assertions that testimonies of personal experiences could be used to definitively establish the empirical reality of spiritualism. He described how his own personal experiences of seeing witnesses being fooled by magic tricks, combined with his empirical investigations with Davey, suggested had led him to believe that human testimony, particularly in the context of magic performances, was much less reliable than past researchers had appreciated. The second part of the Mal-Observation Report, the Experimental Investigation, is authored by Davey. He begins by briefly describing his own personal journey from credulity to skepticism before presenting a series of written statements from the participants, accompanied by his own meta-commentary. Over the course of several months, Hodgson and Davey had invited various members of the public to attend séances hosted by Davey. Most, but not all of the séances took place in Davey's home. All the séances were scripted and choreographed, but they would vary slightly between different audiences, as Davey improvised around individual circumstances and responses. However, at the start of each performance, he would always specifically caution his audience to be mindful of any form of trickery or deception. However, he would not specifically tell them that he was going to be presenting magic tricks, nor did he ever explicitly attribute his performance to supernatural powers. At the conclusion of each demonstration, audience members were asked to provide written accounts in the form of letters, describing everything that they had experienced.

In total, the Mal-Observation Report collected 27 distinct accounts from 17 separate performances. Each account was reproduced in full and was accompanied by commentary from Hodgson and Davey about the ways that the participants' reports deviated from the actual events. Overall, the participants committed consistent and systematic errors in their testimonies: The reports typically omitted crucial elements of the séances—specifically elements that were instrumental for the accomplishment of the conjuring tricks. For example, in one of the scripted segments of the séance Davey would surreptitiously switch a blank slate for another slate that had been pre-marked with writing before the experiment began. Davey conducted most of his performances while seated at a table. He would begin by showing the blank slate clearly; he would then place the slate under the table, where he would switch it for the slate that had writing on it that had previously been concealed beneath the table. He would then remove the slate with writing on it under the table and leave the genuinely blank slate behind. To help conceal the switch, he would not reveal the writing immediately. Instead, he would pretend that the slate was still blank, and explicitly express his disappointment that no spirits had been in contact. This ruse was accepted by his audiences, who remembered examining the original slate. Next, Davey would place the pre-written slate on top of the table (writing side down) and ask the sitter(s) to place their hands upon it. After a brief time, he would then ask them to turn over the slate, revealing the writing. He could then pretend that the writing had just manifested while the participants' hands were on the slate. The trick was only effective because he was able to place the blank slate under the table and replace it with the pre-written one. But the participants' accounts consistently failed to mention that the slate had ever been placed beneath the table at all. These omissions created a powerful illusion of impossibility for the participants, and anyone reading their written accounts would have been unable to reconstruct the true method behind the trick. The trick was deliberately choreographed to maximize the likelihood that participants would either forget or fail to mention the actions that were crucial for accomplishing the illusion. Even a reader well-versed in conjuring and fraudulent slate writing methods would struggle to provide an explanation for how a participant could have examined a slate and then had writing immediately appear while the slate was still in their hands outside the reach of the medium. The Mal-Observation Report's findings suggest that solutions could be found in what was not reported by the participants. This documentation of witnesses' omissions is very similar to the idea of inattentional blindness (Mack & Rock, 1999), a psychological phenomenon that was not formally described and labeled until over a hundred years after the Mal-Observation Report's publication. Davey's actions of placing the slate beneath the table were clearly visible and in plain sight of the participants, and yet their belief that the writing was produced while they held the slates and their omission of his initial placement of the slate under the table indicates that participants nonetheless failed to detect these significant actions. The fact that participants were instructed to watch his actions closely and clearly believed they had done so, suggested that, at least in some instances, their failures to detect the secret methods may have been related to having their attention otherwise engaged. However, given the fact that the reports were collected some time after the critical events, it is not

A Spirit Artist. Sketched by himself, independently, through the mediumship of Fred Evans, upon the surface of one of a pair of slates held in the hands of Mr. and Mrs. J. J. Owen. (*Psychography: Marvelous Manifestations of Psychic Power Given though the Mediumship of Fred P. Evans*, 1893.)

Given through the mediumship of Fred Evans, at a select séance of friends. (*Psychography: Marvelous Manifestations of Psychic Power Given though the Mediumship of Fred P. Evans*, 1893.)

possible to definitively distinguish between failures of awareness that might have occurred during the performance and subsequent failures of memory that might have occurred after the performance.

Beyond documenting omissions in the participants' testimonies, the reports also demonstrated that participants engaged in "positive" errors. Sometimes participants recounted confabulated events that were the products of their own imaginations. For example, one of the participants, Mr. Padshah, described his experience in Sitting IV of the experiment (p. 437-441). According to Padshah, the most remarkable part of his experience was when the written image of his childhood nickname (in Persian), "BOORZU" appeared to manifest on a slate. This was particularly impressive to Padshah, because he had not been called this name in many years, and, to his knowledge, no one else present in the room was aware that he had ever been called that. In a note, Hodgson explained exactly how this had actually happened: Davey had secretly written the word "BOOKS" on the slate to suggest a further test (i.e. a prepared trick) involving books in the room. But when Pashah saw the scrawled word, he mis-read it as "BOORZ" and this minor proofreading error led him to the extraordinary conclusion that a spiritual entity was addressing him personally by a name that no one else in the room, including the medium, could possibly have known by natural means. Davey quickly improvised by wiping the word away before Padshah or anyone else could get a clear look at it, and the other participants were highly impressed by Padshah's dramatic reaction. In general, contemporary researchers and commentators including the Mal-Observation Report's authors, seemed open to the idea that errors of omission, i.e. forgetfulness and inattention, were fairly commonplace; however, the idea that honest, healthy, sober, intelligent witnesses could form reports based on memories of falsely remembered events was highly controversial. Yet the fact that people can and do create false memories of imagined events that can feel subjectively indistinguishable from "real" memories of actual events is exactly what the evidence from Hodgson and Davey's study indicated, and this idea has been borne out by subsequent experimental psychological research that took place many years afterwards (e.g. Bartlett, 1932; Loftus, 2003).

Initial Responses to the Report and the Follow-up Article Regarding Davey's Methods

Reactions to the Mal-Observation Report, following its publication, also indicate robust failures of metacognition in some of its readers. A lively written debate was recorded in the SPR's *Journal* throughout the following years. One of the writers who actively engaged in the discussion around the Mal-Observation Report was Alfred Russell Wallace, who, at the time, was something of a scientific celebrity, having recently co-discovered the theory of evolution with Charles Darwin (Slotten, 2006). While Wallace did not personally attend any of Davey's séances, he attended similar slate writing demonstrations by other mediums including Eglinton. Wallace had been utterly convinced that he had witnessed genuine supernatural phenomena. Even after Hodgson and Davey had explicitly written that they had been using magic trick methods, Wallace remained convinced that the séances conducted by Davey could potentially have involved genuine spiritual phenomena. He wrote an angry letter to the SPR suggesting that Hodgson and Davey were engaged in a deliberate cover-up of Davey's mediumistic talents. Wallace took particular issue with the fact that the original report did not provide full methodological explanations of the conjuring techniques that were used to accomplish the simulated psychography phenomena. He went so far as to state that "unless all [the reported phenomena] can be so explained many of us will be confirmed in our belief that Mr. Davey was really a medium as well as a conjurer, and that in imputing all his performances to 'trick' he was deceiving the Society and the public" (Wallace, 1891, p. 43).

Wallace's critique neatly illustrates the unfortunate potential consequences of actively employing deception as a tool to demonstrate how powerful deceptive methods can be. In another instance, Arthur Conan Doyle, a committed spiritualist, famously maintained that his friend, the magician Harry Houdini, actually accomplished his escapes using psychic powers; all in spite of Houdini's explicit disavowal and active debunking of psychic phenomena (Doyle, 1930). Similar issues continuously haunt those who use magic methods and deceptions as tools to explore belief in paranormal phenomena. Fast forward to the 1970s, Martin Johnson, a psychologist and parapsychologist, wrote-up another example that occurred in 1976, when he arranged for the Swedish magician Ulf Mörling to perform at the 19th Annual Convention of parapsychologists. Mr. Mörling's performance was intended as entertainment, and he was explicitly billed as a magician who would be performing fake demonstrations of psychic abilities. Nonetheless, Johnson wrote that both he and Mörling were taken aback when several of the attending researchers refused to believe that the performance was not a genuine demonstration of psychic phenomena:

> *After the seance, something which I consider as rather embarrassing happened: quite a number of parapsychologists started to discuss the possibility that Mr. Moerling was a medium without being aware of it? More or less fantastic hypotheses were put forward as to how to explain his success... I am positively shocked that so many parapsychologists (between 10 and 20 according to Mr. Moerling's and my independent estimations) in spite of his assurance that his performance was based on trickery, seriously put forward the PSI hypothesis as an explanation for what they had observed.* (Johnson, 1976, p. 4)

Later on, in the 1980s, the magician James Randi orchestrated an elaborate hoax to publicly discredit the scientific rigor of contemporary parapsychological researchers. Randi's methods involved having magicians infiltrate a lab posing as test subjects (Randi, 1983a; 1983b). In his critique of Randi's efforts, sociologist (and magician) Marcello Truzzi even directly referenced Hodgson and Davey's report to help illuminate the issues around using fraudulent methods to expose fraud, noting that, logically, "An irony in such cases is that once the defrauders admit to their fraud, the door is open to distrust them entirely." And I've already mentioned my, even more recent, personal experience of failing to convince students about the fraudulent nature of my own performances at Goldsmiths, University of London in 2019.

Hodgson and Davey were intimately aware of this issue. In the immediate aftermath of Mal-Observation Report's publication, Davey personally resisted publishing all of his methods (Hodgson, 1892).

He told Hodgson that he was concerned that knowledge of the specific methods that they used would potentially lead other investigators to become overly confident when attempting to debunk professional mediums, and, additionally, if people interested in committing future frauds read their methods, then they would be able to adapt their own performances to be even more deceptive. Furthermore, he wished to continue conducting eye-witness experiments using the same methods that had been used in the Mal-Observation Report. Sadly, Davey did not have the opportunity to conduct further research. He died of typhoid in 1891 at the age of 27 ("Obituary," 1891). Shortly after Davey's death, Hodgson acceded to Wallace and other commentators' demands and published a follow-up article, titled "Mr. Davey's Imitations by Conjuring of Phenomena Sometimes Attributed to Spirit Agency" (1891). Here, Hodgson provided many additional details about the methods that Davey had used in the séances, including the choreography of the slate switching, the use of "invisible threads" to seemingly animate bits of chalk, and the fact that Davey had used a small "thimble writer" (a device designed secretly to affix a pencil to his finger) to surreptitiously mark the slates during his performances. He also cited three key texts that contained instructions on how to perform the tricks that Davey had presented. The works referenced included Professor Hoffmann's (a.k.a. Angelo John Lewis's) *Modern Magic* and *More Magic* (Lewis, 1876; 1889), Truesdell's *The Bottom Facts Concerning the Science of Spiritualism* (1883), and the anonymously authored *Revelations of a Spirit Medium* (1891).

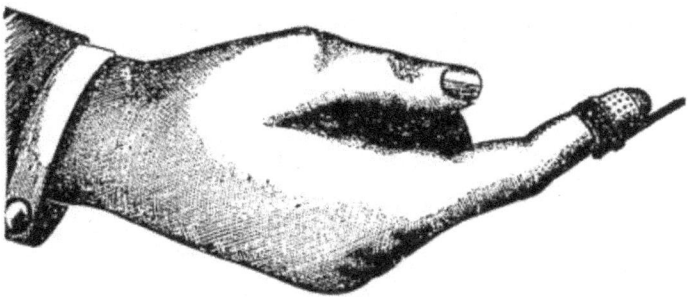

Example of a thimble writer, as used by S. J. Davey.
(*Bottom Facts Concerning the Science of Spiritualism*, 1883.)

Haunted Histories:
Connections Between the Mal-Observation Report and Subsequent Academic Research

As experimental psychology has evolved as a formal academic discipline, researchers have made significant strides in advancing and refining our scientific understanding about the cognitive mechanisms underlying human memory and attention. Subsequent empirical publications have borne out some of the most controversial claims reported by Hodgson and Davey back in 1887. Their work arguably foreshadowed the formal recognition of key theoretical concepts, including inattentional blindness and reconstructive memory. Furthermore, the Mal-Observation Report broadly preempted developments in anomalistic psychology and "the science of magic." Even though, ironically, the Mal-Observation Report has itself been largely overlooked and forgotten by academics and scientists, excepting a few researchers writing about the psychology of magic, parapsychology, and anomalistic psychology (e.g. Truzzi, 1987, Wiseman, Greening, and Smith, 2003; French & Stone, 2017; Tompkins, in press).

Inattentional Blindness

For readers unfamiliar with the term "inattentional blindness," and who may not have seen a demonstration of the now classic basketball game paradigm, I highly recommend that you take a break quick from this text, grab your most convenient internet-connected device, and seek out the YouTube video, posted by Daniel Simons titled: "Selective Attention Test" (Simons, 2010). I'll describe the relevant phenomenon in detail below, but, as with many illusions, an abstract textual description is a poor substitute for experiencing the effect for yourself.

―

Welcome back. I'll get to the weird basketball game in a moment. But first, a bit of context: The concept of inattentional blindness was not formally established in psychological literature until the 1990s, the big break came when psychologists Irving Mack and Arian Rock published a book detailing a groundbreaking set of experiments that explored how participants could fail to notice seemingly obvious stimuli that were fully visible and in plain sight when their attention was otherwise engaged (Rock, Grant, & Mack, 1992; Mack & Rock, 1998). All of Mack and Rock's experiments involved simple computerized displays. And the experiments generally follow a basic pattern. Participants are given a simple primary task, and then, while they are engaged with that task, they are unexpectedly presented with a critical stimulus. For example, in the cross-arm discrimination experiments, the participant's primary task is to look at a series of crosses (+'s) that appear on an otherwise blank screen. Their job is simply to indicate which arm of the cross they think seems longest (either the horizontal or vertical line). It's a simple task, but it nonetheless engages their attention. While they are examining the crosses, an unexpected object (the crucial stimulus), would appear briefly on the screen. The critical stimulus could be a simple shape, or a picture, or a word. After the critical stimulus is presented, the participants are then asked if they had noticed anything else on their screen in addition to the cross. Most participants had no idea that anything else had appeared on the screen, and these folks were considered to be 'inattentionally blind' to the critical stimulus. These results are highly interesting to visual scientists, and the whole book tells a delightful sort of research detective story, with dramatic twists and turns, if your tastes run towards academic prose. Their results also raised a tantalizing follow-up question: If we can demonstrate inattentional blindness for simple static shapes on a computer monitor, how far can we extend these findings into more realistic settings? Mack and Rock's experimental framework set the stage for the basketball game experiment I mentioned earlier: a.k.a. Daniel Simons and Christopher Chabris' Invisible Gorilla Paradigm (Simons & Chabris, 1999; Simons, 2010). Their iconic study took Mack and Rock's ideas and applied them to a more naturalistic and dynamic stimulus set, a live action video recording of two teams of players wearing either white or black t-shirts. For their primary task, participants were instructed to count the number of times players in white passed the basketball, all while trying not to be distracted by the actions of the black-shirted players. The critical stimulus, rather than being a simple shape on the monitor, was a recording of a person wearing a full-body gorilla suit. The gorilla figure is fully visible for 10 seconds. Nonetheless, when people are attending to the basketball passes, more than half will completely fail to notice the gorilla. Subsequent experiments have demonstrated this phenomenon can apply to various stimuli including simple geometric shapes, airplanes, motorcycles, unicycling clowns, fire extinguishers, simulated assaults, and weapons (Haines, 1991; Most & Astur, 2007; Hyman et al., 2010; Castel et al., 2012; Simons & Schlosser 2017).

Inattentional blindness is a highly reliable phenomena; it's not limited to artificial displays, and it is also deeply counter-intuitive (Chabris & Simons, 2010). In surveys designed to explore the general population's beliefs about memory and attention, the majority of respondents tend to endorse the statements that "Human memory works like a video camera, accurately recording the events we see and hear so that we can review and inspect them later" and that "People generally notice when something unexpected enters their field of view, even when they're paying attention to something else" (Simons & Chabris, 2011; 2012). Inattentional blindness studies have shown us that the truth is significantly weirder than most of us tend to imagine. Experimental evidence for inattentional blindness can be found in several studies that pre-date Mack and Rock's terminology. Interestingly, the basic idea behind the basketball game experiment can be traced back to the 1970s, when Ulric Neisser developed a similar experiment involving a video of two teams of basketball passers (e.g. Neisser & Becklen, 1975). Instead of a gorilla costume, his critical stimulus was a woman carrying a large parasol (think Mary Poppins). But Neisser's 'selective looking' experiments failed to achieve the widespread recognition that the invisible gorilla would go on to achieve, arguably because scientists at the time lacked the theoretical framework that Mack and Rock would establish 20 years later. Looking even further back, to the 1880's we can also see evidence of inattentional blindness in the Mal-Observation Report. For Hodgson, inattentional blindness (although not by that name) was one of the most fantastic elements of Davey's performances. Hodgson attended many of Davey's performances as an observer, and he noted in his report that "none of the sitters witnessed that best phenomenon, *Mr. Davey writing* (Hodgson & Davey, 1887, p. 401 original emphasis). Being well acquainted with Davey's methods, Hodgson would not have been surprised by the tricks themselves, but he was consistently amazed at how other, more naive, sitters would reliably fail to notice what was really happening right in front of their eyes.

Today, historical descriptions of inattentional blindness research usually begin with Mack & Rock's seminal experiments. Occasionally, Ulric Neisser's work on selective looking will be mentioned. But I propose that some of the best and earliest experimental evidence for the phenomena can be found even further back in the Mal-Observation Report, published well over a century before the concept acquired mainstream scientific acceptance.

Reconstructive Memory and the Unreliability of Eye-Witness Testimony

There is now a rich psychological literature on the potential unreliability of eye-witness testimony. The concept of reconstructive memory is now widely accepted by the scientific community. The term refers to the idea that our memories are not recordings to be played back as much as narratives that we "reconstruct" each time we engage in recall; as a consequence, what we perceive as memories are not only traces of past events, but are also made up from our expectations, imaginations, and assumptions (e.g. Roediger, 2001). Reconstructive mechanisms of recall can result in accurate memories; however, the process is, by its nature, prone to producing errors and false memories. Like with inattentional blindness, this a deeply counterintuitive concept; the process of reconstruction is very different from the generally accepted intuition that memory operates analogously to a recording device. Experimental evidence demonstrating how easy it is to generate false memories provides the main empirical support for the theoretical idea of reconstructive memory. And researchers have now repeatedly demonstrated that people are capable of confusing imaginary events with real memories (see Loftus, 2005 for a review).

A key breakthrough in this literature can be traced to research conducted in the 1970s on how leading questions can cause witnesses to mistakenly report having seen imaginary events. Loftus and Palmer (1974) demonstrated that participants could be induced to remember seeing events that were not presented in response to leading questions. One week after having watched a video of a car accident, participants were explicitly asked "Did you see any broken glass?" The reported false memories of broken glass could not have been derived directly from the video, because the video did not show any broken glass; thus, the false memory was arguably induced by the question itself, which caused people to imagine broken glass, and confuse those imaginings with their memory of seeing the video footage. These results are particularly important because they demonstrate that leading questions do not merely have the potential to change people's interpretations of their own recollections; they can reliably cause people to create false memories of entirely fictional events. Other, later, studies have demonstrated that false verbal suggestions presented co-currently with events can also induce false reports (Wiseman, Greening, & Smith, 2003; Wiseman & Greening, 2005; Wilson & French, 2013). While researchers continue to explore and debate many details, the theory of reconstructive memory is now widely accepted. But like with inattentional blindness, conventional narratives around the research do not extend back into the 1880s, rather they typically begin with the work of Fredrick Bartlett (1932), who not only drew a clear distinction between reproductive and reconstructive memory, but also argued that reconstructive memory processes were more typical of how people remember events in real-world circumstances. The Mal-Observation Report provided robust evidence of reconstructive memory nearly half a century earlier.

Anomalistic Psychology and The Science of Magic

Empirical investigations involving performance magic and illusion played a small but critical role in the establishment of Experimental Psychology as a scientific discipline (e.g., Wundt, 1879; Pettit, 2013). Several prominent early psychologists wrote about the psychology of magic tricks, particularly as they related to illusions of spiritualist phenomena (Jastrow, 1888, 1896; Dessoir, 1893; Binet, 1894; Triplett, 1900; see also Lamont, 2010; Thomas, Didierjean, & Nicolas, 2017). Most of these early writings represent early examples of a subdiscipline now referred to as anomalistic psychology. This is a sub-discipline of psychology that actively seeks naturalistic explanations for why many people report having extraordinary experiences that seem to be paranormal, or outside the realm of conventional science. Christopher French has written that anomalistic psychology "attempts to explain paranormal and related beliefs and ostensibly paranormal experiences in terms of known (or knowable) psychological or physical factors" (French, 2001, p. 356). The Mal-Observation Report represents an excellent early example of this principle, in that the researchers deliberately used naturalistic methods (i.e. conjuring/magic tricks) to help develop a better psychological understanding of how people might come to report extraordinary paranormal experiences. Notably, the power of these deceptions to falsely induce paranormal experiences has not diminished with time: In 1980, Singer and Benassi used magic trick methods to convince university students that they had witnessed a genuine demonstration of psychokinesis (aka metal bending). More recently, Lesaffre and colleagues (2021), showed that contemporary undergraduate students still reliably endorsed mediumistic explanations in response to magic performances (see also Mohr & Kuhn, 2020). Other researchers have adapted some of the exact same methods used by Davey to create convincing illusions of extraordinary technology, convincing participants that dummy equipment is variously capable of reading their minds or controlling their thoughts (Ali et al., 2014; Olson et al., 2016; 2023); instead of pretending that the source of the extraordinary phenomena is attributable to spirits, these researchers gesture towards (fictional) advanced neuroscientific technology and complex algorithms. These continued investigations promise to help shed new light on how people can formulate and update their beliefs in response to unusual experiences. Looking back on the Mal-Observation Report through the lens of contemporary research in anomalistic psychology, an important takeaway is how relatively minor errors of perception and memory, combined with a bit of deliberate deception and a pre-existing will to believe, can create a cascade of reasoning errors that can result in the formation of extraordinary beliefs.

Performance magic was largely ignored by the scientific community throughout the 20th century (Hyman, 1989; Tompkins, 2024). There is no definitive answer as to why 20th-century psychologists largely disengaged with magic as a research tool; however, two potential reasons include early psychologists' motivations to actively distance themselves from what they viewed as the pseudoscientific associations of psychical research (e.g. Coon, 1992; Sommer, 2012, 2014) and the prominence of behaviorism (e.g. Watson, 1913). Both explanations stem from researchers generally attempting to establish psychology as a formally recognized science. In the case of psychical research, this meant establishing clear boundaries between the scientific psychology and psychical research, and magic tricks were arguably too closely associated with mediumistic fraud for researchers to be comfortable engaging with them professionally. Similarly, behaviorism, with its rejection of any type of psychological data that could not be reduced to an externally measurable behavior effectively rules out the study of magic and illusion, given that these experiences only occur as internal representations in people's minds.

However, throughout the past 25 years, researchers have increasingly been turning to magic trick methods as tools for developing new paradigms to explore human cognition (Lamont & Wiseman, 1999; Kuhn, 2019; Tompkins, 2019). Performance magic represents a rich, and still relatively underutilized resource for cognitive scientists (Thomas, Didierjean, Maquestiaux, & Gygax, 2015). The idea of adapting magic trick methods to psychology experiments has proven to be productive, with researchers developing new laboratory paradigms that help illuminate a wide variety of cognitive mechanisms, including inattentional blindness (e.g., Kuhn & Tatler, 2005; Kuhn & Findlay, 2010, 2011; Barnhart & Goldinger, 2014), change blindness (Smith, Lamont, & Henderson, 2013; Kuhn, Teszka, Tenaw, & Kingstone, 2016), introspection (Johansson, Hall, Sikstrom, & Olson, 2005); decision making (Olson et al., 2015; Shalom, 2013), problem solving (Danek, et al., 2014), memory (Danek, 2013; Wilson & French, 2014), and amodal perception (Beth & Ekroll, 2015; Tompkins, Woods, Aimola Davies, 2016). Although neither Hodgson nor Davey identified as psychologists, the Mal-Observation Report arguably established the first ever framework for how to adapt magic trick methods to aid in psychological research.

Conclusion

Throughout their writing, Hodgson and Davey were careful to acknowledge that just because they were able to simulate mediumistic phenomena using magic trick methods, it did not necessarily follow that every testimony related to a spiritualistic experience was attributable to deception. Both men left open the possibility that scientific advancements and new experimental work might one day vindicate some of the empirical claims of proponents of modern spiritualism. To date, while discussions do continue (Cardeña, 2018; Reber & Alcock, 2020), the mainstream scientific establishment has not come to embrace the principles of modern spiritualism in ways that were foretold by writers like Alfred Russell Wallace and Arthur Conan Doyle. However, the Mal-Observation Report itself has proven remarkably prescient. In light of later scientific developments, the report can be seen as a valuable contribution to psychological science. Many of the report's most controversial findings about testimony, deception, attention, and memory have been borne out by subsequent formal empirical investigations related to inattentional blindness, reconstructive memory, anomalistic psychology, and the science of magic. The following pages contain reproductions of the two key papers that make up the Mal-Observation Report: "The Possibilities of Mal-observation and Lapse of Memory from a Practical Point of View," co-written by Hodgson and Davey and originally published in *The Proceedings of the Society for Psychical Research* in 1887, along with a follow-up paper written by Hodgson titled, "Mr. Davey's Imitations by Conjuring of Phenomena Sometimes Attributed to Spirit Agency," also published in the SPR's *Proceedings*, but four years later, in 1891. Personally, I have yet to encounter compelling evidence that any of us can contact the dead in a mediumistic sense, but this is arguably one of the next best things. Hodgson and Davey's voices live on through their writings, and can still provide valuable lessons for contemporary readers.

References

Ali, S. S., Lifshitz, M., & Raz, A. (2014). Empirical neuroenchantment: From reading minds to thinking critically. *Frontiers in Human Neuroscience, 8*, 357.

Baird, A. (1949). *Richard Hodgson: The story of a psychical researcher and his times, etc*. Psychic Press.

Baker, D. (2019, March). Magic is helping unlock the mysteries of the human brain. *Wired*. https://www.wired.com/story/magic-neuroscience/

Barnhart, A. S., & Goldinger, S. D. (2014). Blinded by magic: Eye-movements reveal the misdirection of attention. *Frontiers in Psychology, 5*, 1461.

Bartlett, F. C. (1932). *Remembering: A study in experimental and social psychology.* Cambridge University Press.

Beard G. M. (1879). The psychology of spiritism. *North American Review 129*, 65-80.

Benassi, V. A., Singer, B., & Reynolds, C. B. (1980). Occult belief: Seeing is believing. *Journal for the Scientific Study of Religion, 19*(3), 337–349.

Beth, T., & Ekroll, V. (2015). The curious influence of timing on the magical experience evoked by conjuring tricks involving false transfer: Decay of amodal object permanence? *Psychological Research, 79*(3), 513–522.

Binet, A. (1896). Psychology of prestidigitation (M. Nichols, Trans.). *Annual Report of the Board of Regents of the Smithsonian Institution.* Washington, DC: Government Printing Office.

Blavatsky, H. P. (1888). *The secret doctrine: The synthesis of science, religion and philosophy.* Theosophical Publishing House.

Brown, G. R. (2017). Henry Slade and his slates: From Europe to the fourth dimension. *Gibecière, 23*, 9–112.

Cadwallader, M. E. (1917). *Hydesville in history.* Progressive Thinker Publishing House.

Capron, E. W. (1850). *Singular revelations: Explanation and history of the mysterious communion with spirits.* Finn & Rockwell.

Cardeña, E. (2018). The experimental evidence for parapsychological phenomena: A review. *American Psychologist, 73*(5), 663–677.

Castel, A. D., Vendetti, M., & Holyoak, K. J. (2012). Fire drill: Inattentional blindness and amnesia for the location of fire extinguishers. *Attention, Perception, & Psychophysics, 74*(7), 1391–1396.

Cattell, J. M. (1898). Early psychological laboratories. *Science, 47*(128), 544–548.

Chabris, C. F., & Simons, D. (2010). *The invisible gorilla: And other ways our intuitions deceive us.* Harper Collins.

Chabris, C. F., Weinberger, A., Fontaine, M., & Simons, D. J. (2011). You do not talk about Fight Club if you do not notice Fight Club: Inattentional blindness for a simulated real-world assault. *i-Perception, 2*(2), 150–153.

Coon, D. (1992). Testing the limits of sense and science: American experimental psychologists combat spiritualism, 1880–1920. *American Psychologist, 47*(2), 143–151.

Coon, D., & Mitterer, J. O. (2013). *Introduction to psychology: Gateways to mind and behavior.* Cengage Learning.

Coues, E. (1892). Independent Slate-writing: A fact of nature. *Religio Philosophical Journal, 2*(40), 628-629.

Crookes, W. (1874). *Researches in the phenomena of spiritualism.* J. Burns.

Danek, A. H., Fraps, T., von Mueller, A., Grothe, B., & Öllinger, M. (2014). Working wonders? Investigating insight with magic tricks. *Cognition, 130*(2), 174–185.

Danek, A. H., Fraps, T., von Müller, A., Grothe, B., & Öllinger, M. (2013). Aha! experiences leave a mark: Facilitated recall of insight solutions. *Psychological Research, 77*(5), 659–669.

Dessoir, M. (1893). The psychology of legerdemain (E. S. Boyer, Trans.). *The Open Court, 7,* 3599–3606.

Doyle, A. C. (1926). *The history of spiritualism.* Cassell & Co.

Doyle, A. C. (1930). *The edge of the unknown.* John Murray.

Eysenck, M. W., & Brysbaert, M. (2018). *Fundamentals of cognition.* Routledge.

Farmer, J. S. (1886). *'Twixt two worlds: A narrative of the life and work of William Eglinton.* Psychological Press.

French, C. C. (2001). Why I study anomalistic psychology. *The Psychologist, 14*(7), 356–357.

French, C. C. (2003). Fantastic memories: The relevance of research into eye-witness testimony and false memories for reports of anomalous experiences. *Journal of Consciousness Studies, 10*(6–7), 153–174.

French, C. C., & Stone, A. (2014). *Anomalistic psychology: Exploring paranormal belief and experience*. Palgrave Macmillan.

Gardner, M. (1992). How Mrs. Piper bamboozled William James. In *Are universes thicker than blackberries?: Discourses on Gödel, magic hexagrams, Little Red Riding Hood, and other mathematical and pseudoscientific topics* (pp. 252–262). Norton.

Haines, R. F. (1991). A breakdown in simultaneous information processing. In *Presbyopia research: From molecular biology to visual adaptation* (pp. 171–175). Springer US.

Hale, N. G. (Ed.). (1971). *James Jackson Putnam and psychoanalysis: Letters between Putnam and Sigmund Freud, Ernest Jones, William James, Sandor Ferenczi, and Morton Prince, 1877–1917*. Harvard University Press.

Harrison, V. (1997). *H. P. Blavatsky and the SPR: An examination of the Hodgson report of 1885*. Theosophical University Press.

Hodgson, R. (1885). Report of the committee appointed to investigate phenomena connected with the Theosophical Society. *Proceedings of the Society for Psychical Research, 3*, 201–400.

Hodgson, R. (1886). On the reports printed in the Journal in June, of sittings with Mr. Eglinton. *Journal of the Society for Psychical Research, 3*, 461–471.

Hodgson, R. (1892). Mr. Davey's imitations by conjuring of phenomena sometimes attributed to spirit agency. *Proceedings of the Society for Psychical Research, 8*, 252–310.

Hodgson, R. (1898). A further record of certain phenomena of trance. *Proceedings of the Society for Psychical Research, 13*, 284–582.

Hodgson, R., & Davey, S. J. (1887). The possibilities of mal-observation and lapse of memory from a practical point of view. *Proceedings of the Society for Psychical Research, 4*, 381–495.

Houdini, H. (1924). *A magician among the spirits*. Harper.

Hyman, I. E., Jr., Boss, S. M., Wise, B. M., McKenzie, K. E., & Caggiano, J. M. (2010). Did you see the unicycling clown? Inattentional blindness while walking and talking on a cell phone. *Applied Cognitive Psychology, 24*(5), 597–607.

Hyman, R. (1989). The psychology of deception. *Annual Review of Psychology, 40*, 133–154.

Irwin, H. J., & Watt, C. A. (2007). *An introduction to parapsychology*. McFarland.

Jastrow, J. (1888). The psychology of deception. *The Popular Science Monthly, 34*, 148–157.

Jastrow, J. (1896). Psychological notes upon sleight-of-hand experts. *Science, 3*(71), 685–689.

Johansson, P., Hall, L., Sikström, S., & Olsson, A. (2005). Failure to detect mismatches between intention and outcome in a simple decision task. *Science, 310*(5745), 116–119.

Johnson, M. (1976). Some reflections after the PA convention. *European Journal of Parapsychology, l*(3). 1-5.

Kuhn, G. (2019). *Experiencing the impossible: The science of magic*. MIT Press.

Kuhn, G., Findlay, J. M. (2010). Misdirection, attention and awareness: Inattentional blindness reveals temporal relationship between eye movements and visual awareness. *Quarterly Journal of Experimental Psychology, 63*(1), 136–146.

Kuhn, G., & Tatler, B. W. (2011). Misdirected by the gap: The relationship between inattentional blindness and attentional misdirection. *Consciousness and Cognition, 20*(2), 432–436.

Kuhn, G., Teszka, R., Tenaw, N., & Kingstone, A. (2016). Don't be fooled! Attentional responses to social cues in a face-to-face and video magic trick reveal greater top-down control for overt than covert attention. *Cognition, 146*, 136–142.

Lamont, P. (2013). *Extraordinary beliefs: A historical approach to a psychological problem*. Cambridge University Press.

Lamont, P., Henderson, J. M., & Smith, T. J. (2010). Where science and magic meet: The illusion of a "science of magic." *Review of General Psychology, 14*(1), 16–21.

Lamont, P., & Wiseman, R. (1999). *Magic in theory: An introduction to the theoretical and psychological elements of conjuring.* University of Hertfordshire Press.

Lesaffre, L., Kuhn, G., Jopp, D. S., Mantzouranis, G., Diouf, C. N., Rochat, D., & Mohr, C. (2021). Talking to the dead in the classroom: How a supposedly psychic event impacts beliefs and feelings. *Psychological Reports, 124*(6), 2427–2452.

Lewis, A. J. (1886). How and what to observe in relation to slate-writing phenomena. *Journal of the Society for Psychical Research, 2,* 362–375.

Lewis, E. E. (1848). *A report of the mysterious noises heard in the house of Mr. John D. Fox: Authenticated by the certificates, and confirmed by the statements of the citizens of that place and vicinity.* Hydesville, Arcadia, Wayne County.

Lodge, O. (1909). *The survival of man: A study in unrecognised human faculty.* Methuen & Company.

Loftus, E. F. (2003). Make-believe memories. *American Psychologist, 58*(11), 867–873.

Loftus, E. F. (2005). Planting misinformation in the human mind: A 30-year investigation of the malleability of memory. *Learning & Memory, 12*(4), 361–366.

Loftus, E. F., & Palmer, J. C. (1974). Reconstruction of automobile destruction: An example of the interaction between language and memory. *Journal of Verbal Learning and Verbal Behavior, 13*(5), 585–589.

Mack, A., & Rock, I. (1998). *Inattentional blindness.* MIT Press.

Macknik, S. L., King, M., Robbins, A., Teller, Thompson, J., & Martinez-Conde, S. (2008). Attention and awareness in stage magic: Turning tricks into research. *Nature Reviews Neuroscience, 9*(11), 871–879.

Marshall, M., & Wendt, R. A. (1980). Wilhelm Wundt, spiritism, and the assumptions of science. In *Wundt studies: A centennial collection* (pp. 158–175). Hogrefe.

Massey, C. C. (1886). The possibility of mal-observation in relation to evidence for the phenomena of spiritualism. *Proceedings of the Society for Psychical Research, 4*, 75–110.

Medium, A. (1891). *Revelations of a spirit medium*. Farrington & Co.

Milner, R. (1996). Charles Darwin and associates, ghostbusters. *Scientific American, 275*(4), 96–101.

Mohr, C., & Kuhn, G. (2020). How stage magic perpetuates magical beliefs. In *Mind reading as a cultural practice* (pp. 93–106). Springer.

Most, S. B., & Astur, R. S. (2007). Feature-based attentional set as a cause of traffic accidents. *Visual Cognition, 15*(2), 125–132.

Münsterberg, H. (1908). *On the witness stand: Essays on psychology and crime*. Doubleday, Page & Company.

Neisser, U., & Becklen, R. (1975). Selective looking: Attending to visually specified events. *Cognitive Psychology, 7*(4), 480–494.

Nickell, J. (2008). A skeleton's tale. *Skeptical Inquirer, 32*(1), 17–20.

Noakes, R. (2019). *Physics and psychics: The occult and the sciences in modern Britain*. Cambridge University Press.

Obituary. (1891). *Journal of the Society for Psychical Research, 5*, 16.

Olson, J. A., Amlani, A. A., Raz, A., & Rensink, R. A. (2015). Influencing choice without awareness. *Consciousness and Cognition, 37*, 225–236.

Olson, J. A., Cyr, M., Artenie, D. Z., Strandberg, T., Hall, L., Tompkins, M. L., Raz, A., & Johansson, P. (2023). Emulating future neurotechnology using magic. *Consciousness and Cognition, 107*, 103450.

Olson, J. A., Landry, M., Appourchaux, K., & Raz, A. (2016). Simulated thought insertion: Influencing the sense of agency using deception and magic. *Consciousness and Cognition, 43*, 11–26.

Oxon, M. A. (W. S. Moses). (1882). *Psychography: A treatise on one of the objective forms of psychic or spiritual phenomena*. Psychological Press Association.

Pankratz, L. (2008). Lessons written with a small gimmick. *Gibecière, 3*(2), 123–152.

Podmore, F. (1902). *Modern spiritualism: A history and a criticism*. Methuen.

Professor Hoffmann (A. J. Lewis). (1876). *Modern magic*. George Routledge & Sons.

Professor Hoffmann (A. J. Lewis). (1889). *More magic*. George Routledge & Sons.

Randi, J. (1983a). The Project Alpha experiment: Part 1. The first two years. *Skeptical Inquirer, 7*(4), 24–33.

Randi, J. (1983b). The Project Alpha experiment: Part 2. Beyond the laboratory. *Skeptical Inquirer, 8*(1), 36–45.

Reber, A. S., & Alcock, J. E. (2020). Searching for the impossible: Parapsychology's elusive quest. *American Psychologist, 75*(3), 391–399.

Rock, I., Linnett, C. M., Grant, P., & Mack, A. (1992). Perception without attention: Results of a new method. *Cognitive Psychology, 24*(4), 502–534.

Roediger, H. L., III. (2001). Psychology of reconstructive memory. In *N. J. Smelser & P. B. Bates (Eds.), International encyclopedia of social and behavioral sciences* (pp. 12844–12849). Pergamon.

Sera-Shriar, E. (2022). *Psychic investigators: Anthropology, modern spiritualism, and credible witnessing in the late Victorian age*. University of Pittsburgh Press.

Shalom, D. E., de Sousa Serro, M. G., Giaconia, M., Martinez, L. M., Rieznik, A., & Sigman, M. (2013). Choosing in freedom or forced to choose? Introspective blindness to psychological forcing in stage magic. *PLoS ONE, 8*(3), e58254.

Sidgwick, E. M. (1886a). The charges against Mr. Eglinton. *Journal of the Society for Psychical Research, 2*, 467–469.

Sidgwick, E. M. (1886b). Results of a personal investigation into the "physical phenomena" of spiritualism, with some critical remarks on the evidence for the genuineness of such phenomena. *Proceedings of the Society for Psychical Research, 4*, 45–74.

Simons, D. J. (2010). Monkeying around with the gorillas in our midst: Familiarity with an inattentional-blindness task does not improve the detection of unexpected events. *i-Perception, 1*(1), 3–6.

Simons, D. J., & Chabris, C. F. (1999). Gorillas in our midst: Sustained inattentional blindness for dynamic events. *Perception, 28*(9), 1059–1074.

Simons, D. J., & Chabris, C. F. (2011). What people believe about how memory works: A representative survey of the US population. *PLoS ONE, 6*(8), e22757.

Simons, D. J., & Chabris, C. F. (2012). Common (mis)beliefs about memory: A replication and comparison of telephone and Mechanical Turk survey methods. *PLoS ONE, 7*(12), e51876.

Simons, D. J., & Schlosser, M. D. (2017). Inattentional blindness for a gun during a simulated police vehicle stop. *Cognitive Research: Principles and Implications, 2*(1), 1–8.

Slotten, R. A. (2004). *The heretic in Darwin's court: The life of Alfred Russel Wallace.* Columbia University Press.

Smith, T. J., Lamont, P., & Henderson, J. M. (2013). Change blindness in a dynamic scene due to endogenous override of exogenous attentional cues. *Perception, 42*(8), 884–886.

Sommer, A. (2012). Psychical research and the origins of American psychology: Hugo Münsterberg, William James and Eusapia Palladino. *History of the Human Sciences, 25*(2), 23–44.

Sommer, A. (2014). Psychical research in the history and philosophy of science: An introduction and review. *Studies in History and Philosophy of Science Part C: Studies in History and Philosophy of Biological and Biomedical Sciences, 48*, 38–45.

Stewart, B. (1885). Presidential address. *Proceedings of the Society for Psychical Research, 3*, 64–68.

Thomas, C., Didierjean, A., Maquestiaux, F., & Gygax, P. (2015). Does magic offer a cryptozoology ground for psychology? *Review of General Psychology, 19*(2), 117–128.

Tompkins, M. L. (In press). Haunted Histories of Experimental Psychology: A Re-Examination of Hodgson and Davey's Mal-Observation Report. *Advances in Nineteenth-Century Research*.

Tompkins, M. L. (2017, June). He blinded us with séance. *New Scientist*, *234*(3123), 42–43.

Tompkins, M. L. (2019). *The spectacle of illusion: Magic, the paranormal, and the complicity of the mind*. Thames & Hudson.

Tompkins, M. L. (2024, November 7). A science of magic bibliography: 2024 update. https://www.matt-tompkins.com/blog/2024/11/7/a-science-of-magic-bibliography-2024-update

Tompkins, M. L., Woods, A. T., & Aimola Davies, A. (2016). The phantom vanish magic trick: Investigating the disappearance of a non-existent object in a dynamic scene. *Frontiers in Psychology*, *7*, 950.

Treitel, C. (2004). *A science for the soul: Occultism and the genesis of the German modern*. Johns Hopkins University Press.

Triplett, N. (1900). The psychology of conjuring deceptions. *The American Journal of Psychology*, *11*(4), 439–510.

Truesdell, J. W. (1884). *The bottom facts concerning the science of spiritualism: Derived from careful investigations covering a period of 25 years*. G. W. Carleton & Company.

Truzzi, M. (1987). Reflections on Project Alpha: Scientific experiment or conjuror's illusion. *Zetetic Scholar*, *12*(13), 73–98.

Wallace, A. R. (1891). Correspondence. *Journal of the Society for Psychical Research*, *5*, 43.

Wallace, A. R. (1896). *Miracles and modern spiritualism*. G. Redway.

Watson, J. B. (1913). Psychology as the behaviorist views it. *Psychological Review*, *20*(2), 158–177.

Weisberg, B. (2009). *Talking to the dead: Kate and Maggie Fox and the rise of spiritualism*. Zondervan.

Wilson, K., & French, C. C. (2014). Magic and memory: Using conjuring to explore the effects of suggestion, social influence, and paranormal belief on eyewitness testimony for an ostensibly paranormal event. *Frontiers in Psychology, 5*, 1289.

Wiseman, R. (2025). Letter: S. J. Davey: New information about a pioneering parapsychologist. *Journal of the Society for Psychical Research, 89*, 142–147.

Wiseman, R., & Greening, E. (2005). It's still bending: Verbal suggestion and alleged psychokinetic ability. *British Journal of Psychology, 96*(1), 115–127.

Wiseman, R., Greening, E., & Smith, M. (2003). Belief in the paranormal and suggestion in the séance room. *British Journal of Psychology, 94*(3), 285–297.

THE POSSIBILITIES OF MAL-OBSERVATION AND LAPSE OF MEMORY FROM A PRACTICAL POINT OF VIEW.[1]

INTRODUCTION.
By Richard Hodgson

Concerning the physical phenomena of Spiritualism,[2] Mr. A. R. Wallace has said:—

They have all, or nearly all, been before the world for 20 years; the theories and explanations of reviewers and critics do not touch them, or in any way satisfy any sane man who has repeatedly witnessed them; they have been tested and examined by sceptics of every grade of incredulity, men in way qualified to detect imposture or to discover natural causes— trained physicists, medical men, lawyers, and men of business—but in every case the investigators have either retired baffled, or become converts. (*Miracles and Modern Spiritualism*, pp. 202, 203.)

It has indeed been considered by perhaps the majority of Spiritualists, not only that the recorded testimony to these physical phenomena is enough to establish their genuineness, but that any honest investigator might establish their genuineness to his own satisfaction by personal experience. I agreed in a great measure with this opinion

---

[1] Parts both of Mr. Davey's article and of Mr. Hodgson's introduction have appeared in the *Journal* of the S.P.R.
[2] See *Proceedings*, Part X., p. 45.

when, some ten years ago, I attended my first seance; but hitherto my personal experiences, though not by any means extensive, have been almost precisely of the same nature as Mrs. Sidgwick's (*Proceedings*, Part X., pp. 45, 46); the physical phenomena which I have witnessed were either clearly ascertained by my friends and myself to be fraudulent, or they were inconclusive and accompanied by circumstances which strongly suggested trickery. I regarded this result merely as negative, since I had learnt early in my investigation that spurious manifestations were undoubtedly often produced by professed mediums; and three years ago I was still under the impression that a large mass of reliable testimony existed, adequate to establish the genuineness of at least some of the commoner forms of physical phenomena, and especially of "psychography,"—that is writing without any operation of the medium's muscles. This was also quite recently the opinion of Mr. C. C. Massey, who says (*Proceedings*, Part X. p. 98):—

But original research is not necessary in the first instance. Many, of whom I am one, are of an opinion that the cue for these phenomena generally, and for "autography"[3] in particular, is already complete.

I have long since concluded that I estimated this testimony much too highly. When, in June, 1884, after reading some accounts of "psychography," I had a sitting with Eglinton for "slate-writing," I fully expected to witness phenomena that should be as indubitably beyond the suggestion of trickery as those appeared to be of which I had read and heard descriptions. Writing was produced at my first "slate-writing" sitting with Eglinton, to which I was accompanied by Mr. R. W. Hogg; but Mr. Hogg and myself were both independently of opinion that Eglinton produced the writing himself without the intervention of any extraordinary agency. In writing our detailed report of the sitting, we appreciated, as we had never done before, the difficulties of observation and of recollection, difficulties which we thought must almost effectually prevent a full and accurate description from being given of events analogous to those which we attempted to record. Our report, and the reports of various other sitters with Eglinton, most of whom were, however, convinced of the genuineness of Eglinton's phenome-

---

[3] A word proposed as a substitute for "psychography."

Introduction

na, were printed in the *Journal* of the S.P.R. for June, 1886, with some explanations and criticisms by Mrs. Sidgwick, who drew attention to two of the incidents in his career "which show that we must not assume any disinclination on his part to pass off conjuring performances as occult phenomena." Mrs. Sidgwick, who had previously had the advantage of witnessing some of Mr. Davey's performances, and comparing her reports with those of another witness, and who had therefore been able to form some practical estimate of the frailty of human perception and memory under the peculiar circumstances involved, expressed her opinion that the phenomena recorded in the accounts as having occurred in the presence of Eglinton were attributable to "clever conjuring." In the meantime, in the course of a visit to India for the purpose of investigating the "Theosophical" phenomena of Madame Blavatsky, I had had a somewhat considerable and varied experience in comparing the testimonies of numerous *bona fide* witnesses to events belonging to the class of conjuring performances. The most instructive to me in the first instance were the different accounts which I heard from eye-witnesses of the tricks of the Hindoo jugglers. I saw many of these performances, and saw them frequently, and having learnt secretly from the jugglers themselves how they were done, I was thereafter in a position to compare the accounts of them with the actual occurrences, and I was surprised exceedingly to find to what extent they were misdescribed by intelligent spectators who were unaware of the *modus operandi* of the tricks. With the advantage of this experience, I studied in minute detail the testimony to Eglinton's phenomena recorded in the *Journal* for June, and found that if only the same kinds of misdescription were allowed for in these reports as I had known to be honestly displayed by equally intelligent witnesses, the phenomena were perfectly explicable by conjuring. Not only was this the case, but there were many little incidents mentioned, for the most part innocently and almost casually, in the reports, which afforded indications that if Eglinton's performances were not conjuring, they were very curiously adapted to resemble conjuring operations. And when to these facts, besides the clear evidence of Eglinton's previous imposture, was added the further fact, emphatically pointed out by Mrs. Sidgwick, that every experiment with Eglinton so devised and carried out as apparently to exclude the possibilities of trickery by dispensing with the necessity for continuous observation, had *failed*,[4]—there could, I thought, be little doubt, in the minds of rational and impartial inquirers, of the justice of the conclusion which Mrs. Sidgwick had reached.

But this was not the case. Correspondence and controversy made it clear that the ordinary reader did hesitate to agree with Mrs. Sidgwick, and it soon became manifest that a common but erroneous assumption prevailed concerning the reliability of human testimony under the peculiar circumstances at issue. It appeared that a large number of the readers of Mrs. Sidgwick's article in the *Journal* for June were prejudiced in favour of ordinary human powers of observation and recollection under—it is to be remembered—exceptionally adverse circumstances; and that they were thus prejudiced simply because they had never made any special experiments, with the view of ascertaining exactly how much reliance could be placed upon the reports of even acute and intelligent observers of the "slate-writing" performances of a conjurer known as such. They had decided *a priori* as to the capacity of human perception and memory under quite peculiar conditions, and most of them, I venture to say, had thus decided, not only without possessing any familiarity with the various modes of producing" slate writing" by conjuring, but without possessing any familiarity with conjuring tricks in general, and without being aware of the extent to which we are all subject to *illusions of Memory*, which, in relation to the reports of "psychography," are more deserving of consideration than even illusions of Perception.

It seemed desirable, therefore, to carry out a somewhat more systematic investigation than had heretofore been attempted,—to provide the ordinary reader with the opportunity of comparing for himself the records given of conjuring performances by the uninitiated, with the testimony offered for the genuineness of mediumistic phenomena. To the accomplishment of this task Mr. S. J. Davey has given much valuable labour, as the sequel abundantly shows. This, however, was not enough. It is obvious, of course, that any report is worthless for proving occult agency if a similar report by an equally competent witness is

---

[4] For the discussion of this point see *Proceedings*, X., pp. 70-2, *Journal* for June, 1886, pp. 332, 333, *Journal* for November, 1886, pp. 458-60, *Journal* for December, 1886, pp 475 and 481-85.

given of what is known to be a conjuring trick. But this in itself would not enable us to estimate the true worth of testimony in such cases; on the contrary, it might just as well lead to a new and irrational faith in the unlimited capacity of conjuring. No doubt there are special "dodges," unique wonders of workmanship, staggering flashes of well-nigh incredible dexterity, for which we must always leave ample margin in any pronouncement upon the limitations of a conjurer. Still, with all this, and after the largest allowances have been made for the possibilities of simple failure to observe, it will be admitted that there are numerous records of "psychographic" phenomena that have occurred with mediums (and also with Mr. Davey), which, *as described*, are inexplicable by trickery. It was of the utmost importance, therefore, to determine how far such records might be misdescriptions, and what were the chief causes of the misdescriptions. In the course of a paper contributed to the *Journal* I urged that the principal cause of misdescription, apart from mal-observation, was the untrustworthiness of memory, and I endeavoured to classify roughly the main forms into which the errors of recollection fell. That students of mental science like Mr. Roden Noel and Mr. Massey should put aside so easily the considerations which I alleged with respect to the treachery of memory, suggests that these considerations had, in all probability, been absolutely unheeded by the ordinary recorder, unfamiliar with the more delicate processes of introspective discrimination. And although in the *Journal* I felt it almost needful to apologise for my exposition of the lapses to which we are all liable, on the ground that they had been "almost entirely overlooked by the antagonists of Mrs. Sidgwick's view," the result has shown, not only that they had been entirely overlooked in the degree to which I urged them, but that the most eminent defenders of mediumistic phenomena refused to admit their validity or their significance. But my warrant for the importance of these considerations was much more than the experience of my own lapses in recording, confirmed as that was by the discovery of radical discrepancies between independent reports of one and the same séance, in the cases where such independent reports were given of sittings with Eglinton; nor was it restricted to the lessons which I bad learnt by a comparison of oral and written accounts of common conjuring tricks with my knowledge of the real events; my warrant consisted further in the fact that all the forms of error to which I alluded are actually embodied in the reports of Mr. Davey's performances. In repeating these considerations here, then, I desire the reader to bear in mind that they are not vague theoretical speculations as to possibilities which have rarely if ever been realised, but warnings against veritable pitfalls which are dangerous even to the most wary investigator, into which Mr. Davey's sitters demonstrably fell, and Eglinton's sitters also demonstrably, in the cases which admitted of direct ascertainment.

I shall first recount an incident which occurred in connection with a Hindoo juggler's performance unconnected with Spiritualism, and which produced a deep impression upon myself at the time.

The juggler was sitting upon the ground immediately in front of the hotel, with his feet crossed. Two small carved wooden figures were resting on the ground, about two feet distant from the juggler. Some coins were also lying on the ground near the figures. The juggler began talking to the figures, which moved at intervals, bowing, "kissing," and bumping against each other. The coins also began to move, and one of them apparently sprang from the ground and struck one of the figures. An officer and his wife, who had but recently arrived at the hotel, were spectators with myself, and we stood probably within two yards' distance of the juggler. I knew how the trick was performed; they did not know. The officer drew a coin from his pocket, and asked the juggler if this coin would also jump. The juggler replied in the affirmative, and the coin was then placed near the others on the ground, after which it betrayed the same propensity to gymnastic feats as the juggler's own coins. Two or three other travellers were present at dinner in the evening of the same day, and in the course of the conversation the officer described the marvellous trick which he had witnessed in the afternoon. Referring to the movements of the coins, he said that he had taken a coin from his own pocket and placed it on the ground himself, yet that this coin had indulged in the same freaks as the other coins. His wife ventured to suggest that the juggler bad taken the coin and placed it on the ground, but the officer was emphatic ill repeating his statement, and appealed to me for confirmation. He was, however, mistaken. I had watched the transaction with special curiosity, as I knew what was necessary for the performance of the trick. The officer had apparently intended to place the coin upon the ground himself, but as he was doing so, the juggler leant slightly forward, dexterously and in a most unobtrusive manner received the coin from the fingers of the officer as the latter was stooping down, and laid it close to the others. If the juggler had not thus taken the coin, but had allowed the officer himself to place it on the ground, the trick, as actually performed, would have been frustrated.

Now I think it highly improbable that the movement of the juggler entirely escaped the perception of the officer-highly improbable, that is to say, that the officer was absolutely unaware of the juggler's action at the moment of its happening; but I suppose that although an impression was made upon his consciousness, it was so slight as to be speedily effaced by the officer's *imagination* of himself as stooping and placing the coin upon the ground. The officer, I may say, had obtained no insight into the *modus operandi* of the trick, and his fundamental misrepresentation of the only patent occurrence that might have given him a clue to its performance debarred him completely from afterwards, in reflection, arriving at any explanation.

Just similarly, many an honest witness may have described himself as having placed one slate upon another at a sitting with a "medium," whereas it was the medium who did so, and who possibly effected at the same time one or two other operations altogether unnoticed by the witness.[5]

Now it is the universal mental weakness of which the above incident is an illustration, that forms one of the main sources of error in the reports of "psychography." There are, of course, other sources of error, such as the direct illusions of perception caused by mechanical contrivances or the dexterity of the medium or the dominant expectations of the witness; there is also notably the distraction of the sitter's attention to such an extent that he is not aware at all of certain actions performed by the medium, but this often results in positive misdescription owing to the weakness of memory; as Mrs. Sidgwick remarks (*Journal* for June, 1886), we are liable not only to allow our attention to be distracted, but to forget immediately that it has been distracted, or that the event which distracted it ever occurred"; and the source of error which I desire in particular to press upon the reader's notice is the perishability, the exceeding transience, the fading feebleness, the evanescence beyond recall, of certain impressions which nevertheless

---

[5] For an example see SITTING II, *Note* 7, p. 194.

did enter the domain of consciousness, and did in their due place form part of the stream of impetuous waking thought.

It is, moreover, not simply and merely that many events, which did obtain at the sitting some share of perception, thus lapse completely from the realm of ordinary recollection. The consequence may indeed be that we meet with a blank or a chaos in traversing the particular field of remembrance from which the events have lapsed;[6] but this will often be filled by some conjectured events which rapidly become attached to the adjacent parts, and form, in conjunction with them, a consolidated but fallacious fragment in memory.[7] On the other hand, the consequence may be that the edges of the *lacunæ* close up—events originally separated by a considerable interval are now *remembered* vividly in immediate juxtaposition, and there is no trace of the piecing.[8]

Another source of error which bears a kinship to this depends sometimes upon the absence of a prolonged carefulness in writing out the original record of the sitting. Events which occurred during the sitting, which made a comparatively deep impression, which had not, at the time of recording, sunk beyond the possibility of recall, nevertheless do not appear in the report, because they were temporarily forgotten; and having been thus omitted, the temporary forgetfulness is likely to become permanent, owing to the very coherence given to the defective account by the recording.[9]

Last September I spent many hours recalling and writing notes of a slate-writing séance. The task occupied me some six or seven continuous hours on each of the two days following the evening of the séance. Taking the first page of my Mss., I find, among what are plainly interpolations[10] after the page was originally completed, an

---

[6] *Partial omission.* For an example see SITTING IV, Report I, Mr. Padshah's discussion of [*f*], p. 127.

[7] *Substitution.* For an example see SITTING III, *Note* 5, p. 196.

[8] *Complete omission.* For an example see SITTING II, *Note* 17, p. 195.

[9] *Complete omission temporary.* For an example see SITTING IV, *Note* 12, p.198.

[10] I still recollect, as I think, my surprise at finding, while I was engaged in making the record, that I had forgotten at the moment such an important incident as that referred to in the interpolation; but apart from this, the passage was undoubtedly written afterwards, as appears from its position, &c. I may add that I had probably spent an hour or two in originally noting the first page.

exceedingly noteworthy passage.

I had held the slate against the table instantaneously after the "conjurer" had placed it in position; the slate was shortly afterwards withdrawn, and the chalk which had been placed upon it was found crushed. The chalk marks were cleaned off. A second time I held it similarly, and on withdrawal a dash was found on the slate, which was again cleaned. After noting these and other directly connected events, I had originally written, placing the occurrence before the production of writing: "He then turned the slate over, and put the nib of chalk on, and asked me to hold." My alteration of this reads:

"After holding some time, he asked me to put my holding hand upon his other holding B.'s, so as to complete circuit. With this exception I held the slate in each case against the table. Later, he asked me to hold again. " I had nearly omitted this most important exceptional circumstance here described, correctly described as I have since learnt from the "conjurer." I may further notice that it occurred *before* the first writing was obtained, as I rightly placed it. The "conjurer" did turn the slate over as I originally wrote, on three subsequent occasions during the sitting, but he did *not* do so previous to the appearance of the first writing. My temporary forgetfulness thus involved the temporary insertion of a conjectured event. Or, since the event thus inserted did actually occur later in the sitting, the insertion of it in the wrong place may be regarded as an illustration of the tendency to transposition, to which Mr. Angelo J. Lewis has also drawn attention (*Journal* for August, p. 362), in referring to the difficulty of recalling in their proper order such events as those in question; it is almost impossible to avoid confusing the sequence if the events are crowded, even if they appeared at the time of their occurrence to be of special importance.[11]

In addition to the mistakes which thus originate from the lapsing of certain events beyond recollection, there is the further mistake to which we are liable, of unwittingly inserting events between others which occurred in immediate sequence. This of course also depends upon the weakness of memory; the events as they originally occurred

---

11 *Transposition*. For an example see SITTING XV, *Note* 3, p. 202.

may have acquired only a loose coherence in consciousness, so that an event afterwards *imagined* usurps easily a place in the series and becomes fixed by recording and repetition. A perfectly pure interpolation, that is, one which does not involve either *substitution* or *transposition*, probably does not occur very often, and it would not be easy to establish the fact of its occurrence in any particular case; mixed interpolations are not uncommon.[12]

Now it is quite impossible to estimate rightly the reports of "psychography" and analogous performances without having some experimental knowledge as to how far such reports may be rendered untrustworthy by these faults of partial and of complete omission, of substitution, of transposition, and of interpolation.[13]

Suppose that we are considering the testimony of a witness to his own separate and complete examination of a slate immediately previous to the apparent production of writing. Then, according to what I have been saying, we have—with a perfectly *bona fide* witness—four possibilities to consider besides the one that his impression is correct. It may actually be that no examination at all was made by the witness (interpolation); it may be that, although made, the examination was not made in the perfect manner now described (substitution); it may be that the examination, although faultless and made at the sitting, was not made on the occasion alleged (transposition); or it may be that although the examination was made as described, and on the occasion alleged, events, perhaps unnoticed or regarded by the witness as insignificant, intervened between the examination and the apparent production of the writing (omission).

I need hardly say that in relation to these inherent faults of memory leading to misdescription, we must consider the natural tendency to exaggerate in recording phenomena suggestive of occult agency; hence, in many cases, further omissions and interpolations. But we must carefully distinguish this tendency to exaggerate from another

---

[12] *Interpolation.* For examples see SITTING III [c], p. 122, and *Note* 5, p. 196; SITTING IV, *Note* 15, p. 199; SITTING VIII, *Note* 1, p. 196.

[13] I have not attempted to arrange the faults of memory which I have briefly specified in any system of exclusive division, but rather to exhibit them in their modes of genesis. This is not the place to discuss them in greater detail.

Introduction

cause of transfiguration which affects both perception and memory. I refer to the mental attitude of the sitter during the séance.[14] Events that under ordinary circumstances, or if the witnesses were intent upon discovering a trick, would make a comparatively deep and lasting impression upon consciousness, glide past or are swiftly forgotten, simply because of the absorption of the spectator's interest in the supposed "supernormal" manifestations. The distortion traceable in many reports is largely due to special lapses of memory for the explanation of which we must look chiefly to this peculiar emotional state. I shall refer to this consideration in pointing out some of the difficulties in the way of Mr. Davey's investigations (see p. 69); and its importance, even as regards mal-observation, has, I am sure, been widely under-estimated. We cannot doubt that many a Spiritualist has found his convictions confirmed at some séance by displays of the most paltry imposture, who would, had he attended the séance under the assurance that he was about to witness a conjuring performance, have detected the *modus operandi* instantly. I may give an instance which came under my own observation. At a materialisation séance given by Firman, at which I was present, a supposed "spirit-form" appeared, draped in a semi-transparent flowing robe,—so transparent, in fact, that Firman's bare arm was visible behind it, waving it to and fro. When the figure retired to the "cabinet," the door closed upon a portion of the robe. The door opened again slightly, and the end of the robe was drawn into the "cabinet." Most of the sitters perceived this clearly, but one, a "believer," averred conscientiously that the fabric was not withdrawn, and that he saw it slowly melt away.

I think it will hardly be denied that there are extreme cases where unquestioning faith incapacitates an otherwise intelligent witness; and although I do not challenge Mr. Massey's opinion that for mediumistic phenomena a certain psychical co-operation with the medium on the part of the investigator may be necessary, I entertain no doubt that the witness who gives this co-operation is less likely to discover trickery than the man who is bent on discerning the *modus operandi* of what he knows to be conjuring. Nay, I shall go further and say that there is

---

[14] An illustration of these remarks will be found in the accounts of the different witnesses of SITTING II. See p. 107.

an important difference between the investigator, however "sceptical," who thinks it *possible* that supernormal manifestations will occur, and the investigator who has some solid ground of assurance, independent of his own "scepticism," that the manifestations will be conjuring tricks. The failure to understand this has caused some Spiritualists to put forward the claim that Mr. Davey should produce in their presence a phenomenon similar to, and under the same conditions as, some phenomenon which they describe themselves as having witnessed with a professional medium, or that he should at all events produce the *appearance* of those conditions, &c. In the first place, those who have put forward this claim do not seem to have taken the obvious course of demanding first that the medium should reproduce the phenomenon as they desire Mr. Davey to reproduce it; and in the second place they ignore the fact that their own "psychical condition," so different in the two cases—in one, a favouring co-operation, in the other, a resolve to expose—might be a bar to Mr. Davey, but an open door to the medium. Mr. Massey thinks that the sort of co-operation required is "a mental disposition perfectly consistent with the most scientific vigilance." (*Proceedings*, Part X., p. 98.) But his belief in the possibility of this is not supported by what we know of mental action, since if any attention is being given to a favouring co-operation, probably less will be available in consequence for so dissimilar a task as the exercise of "scientific vigilance"; it is certainly not in accordance with my own experience as a psychically co-operating sitter with mediums, and is directly opposed to the conclusions which I have formed as an observer of Mr. Davey's witnesses.

In estimating, therefore, the value of the testimony to "psychography," we must ask—How much misdescription are we likely to find in the record that maybe due to the ignorance of the witness concerning the points worth mentioning? How much misdescription are we likely to find owing to the impulse of the witness to exaggerate, possibly stimulated by the impetus of a new enthusiasm or the momentum of a cherished belief? We then come to the events as *recollected at the time of the record*. How much distortion from the events as originally perceived must we expect to find in this recollection, owing to the inherent weakness of memory, increased as that may be by the peculiar mental attitude or emotional state of the witness? How far, in the next place,

Introduction

may a description of the events as perceived by the witness, have differed from a full and accurated escription of the events as they actually occurred, owing to the special mal-observation displayed by the witness in consequence of a peculiar mental attitude or emotional state? Finally, how much must we allow for the mal-observation that may be caused by the exceptionally disturbing influence of "a person skilled in particular forms of deception, whose chief object is to prevent the witnesses from perceiving many of the actual occurrences, and to persuade them, by ingenious illusions, to an erroneous belief concerning others"? (*Journal* of the S.P.R. for January, 1887.)

I endeavoured to show in the *Journal* (October, November, Supplement to December, 1886)—making certain assumptions as to the defects of perception and memory under the special circumstances involved that the reports, printed in the *Journal* for June, 1886, of Eglinton's performances, were worthless for proving occult agency. The assumptions which I made are completely justified by the mistakes which have been exhibited by those witnesses of Mr. Davey's performances who are known to myself—persons whose general intelligence knowledge of conjuring, powers of observation and retentiveness, &c.—so far as I can judge of these entitle them to be placed on the same level as the writers of the reports printed in the *Journal* for June, the majority of whom I also know personally. And I have not yet seen any report of "psychography" which, when due allowance is made for the untrustworthiness of observation and recollection, excludes the possibilities of conjuring.

By way of illustration I shall here deal briefly with a case concerning which Mr. C. C. Massey specially challenges judgment. It is true that he does this chiefly upon the question of mal-observation, but the case will serve as well to illustrate the possibilities of memory illusion and trick mechanism. Mr. Massey quotes from the case in *Proceedings*, Part X., pp. 87, 88, and refers to his report in *Light* of April 19th, 1884, from which I take the following extract:—

There was a pile of Mr. Eglinton's own slates upon the table, and it was always upon one or other of these that the writing was obtained. Of the two that were used, I cleaned one, after it had been well wetted, with a dry sponge, myself, on both sides; the other I saw similarly treated by Mr. Eglinton. Of course I watched to see that there was no unobserved change of slate, nor did Mr. Eglinton rise from his seat during the séance, except once, to write down an address I had given him. It will be understood that we sat in broad daylight.

We noticed two facts (always observed likewise with Slade), one of which, certainly, could not result from any voluntary act of the medium. This was the lowering of the temperature of the hand which held the slate, just before and after the writing. The other fact was the cessation of the sound of writing when Eglinton broke the contact of his hand with my own.

From my experience with Slade, I was sure that success was near when I felt the coldness of the medium's hand, as he rested it, with the slate, on the table, just before the writing came. Mr. Eglinton now laid one of the two equal-sized slates (10 ¾ inches by 7 ⅝) flat upon the other, the usual scrap of pencil being enclosed. Both slates were then, as I carefully assured myself, perfectly clean on both surfaces. He then forthwith, and without any previous dealing with them, presented one end of the two slates, held together by himself at the other end, for me to hold with my left hand, on which he placed his own right. I clasped the slates, my thumb on the frame of the upper one (⅞ inch), and three of my fingers, reaching about four inches, forcing up the lower slate against the upper one. We did not hold the slates underneath the table, but at the side, a little below the level. Mr. Noel was thus able to observe the position. Mr. Eglinton held the slates firmly together at his end, as I can assert, because I particularly observed that there was no gap at his end. I also noticed his thumb on the top of the slates, and can say that it rested quite quietly throughout the writing, which we heard almost immediately and continuously, except when Mr. Eglinton once raised his hand from mine, when the sound ceased till contact was resumed.

The inner surface of one of the slates was shortly afterwards found covered with writing.

Mr. Roden Noel corroborates Mr. Massey's description, saying "Every word of this account I am able to endorse."

Now I suppose that the writing had been prepared by Eglinton beforehand, and that it was upon one of the slates which Mr. Massey was then holding. How much (a) mal-observation, or (b) lapse of memory, or (c) ignorance of conjuring contrivances on the part of Mr. Massey does this supposition appear to involve?

Introduction

(a) It must be observed, to begin with, that the phenomenon was not a simple and isolated one; nor was it, so far as appears from the account, suggested by Mr. Massey, or previously prepared for by him; the slates were Eglinton's, and there was a pile of Eglinton's slates on the table; Mr. Massey's attention, moreover, seems to have been partly given to the temperature of Eglinton's hand.

What Mr. Massey really meant when he wrote in *Light*, "Both slates were then, as I carefully assured myself, perfectly clean on both surfaces," is by no means clear.[15] *When* did Mr. Massey assure himself? before or after Eglinton laid one slate upon the other? If after, are we to presume that he took the slates into his own hands and examined all four surfaces? Who, in this case, placed the slates together again? Mr. Massey, or Eglinton?—"we must have particularity of statement, evidence that the witness has himself analysed the observation into the acts of perception constituting it, and that at the time of observation." (*Proceedings*, Part X., p.89.) We should, I think, do least violence to Mr. Massey's report if we suppose him to have meant to say that he had carefully assured himself just *before* Eglinton laid one slate upon the other; and in this case I hold that Mr. Massey's observation could have been deceived, that there might have been one side of one slate, which he never saw, or that another slate might have been substituted for one of two slates, both sides of which he did see. Mr. Massey does not state that he took the two slates in question into his own hands, and I have no ground for supposing that at that time he was an expert in detecting sleight-of-hand manipulations of slates.[16] So much for the amount of mal-observation required.

(b) Proceeding now to lapse of memory, let us suppose that the slates were clean when Eglinton laid them together as described. Mr.

---

[15] The above sentence (together with Mr. Noel's endorsement) hardly seems exactly equivalent to "our statement that Mr. Eglinton, after enclosing the pencil within the slates which we *then* carefully assured ourselves were both quite clean on both surfaces," &c. (*Proceedings, loc. cit.*). There is nothing said in the original report about *Mr. Noel's* "carefully assuring" himself, and the meaning of then is here less apparently ambiguous than in the original report.

[16] To avoid complication I am dealing with points of observation and memory as separately as possible. But in connection with Mr. Massey's statement which I have considered above, I might have questioned whether he had not misplaced his feeling of assurance or the process by which he assured himself.

Massey says that Eglinton "then forthwith, and without any previous dealing with them, presented," &c. The sitting was in the afternoon, and Mr. Massey wrote his account of it in the evening of the same day. We have to consider therefore whether it is possible that Mr. Massey —with the grant of an exceptionally good memory—should have *remembered* Eglinton's presentation of the slates to him as having immediately followed upon Eglinton's original placing of them together, although these events might in reality have been separated by an interval during which Eglinton might have changed one of the slates for a third. For example, suppose that when Eglinton lifted the slates the pencil dropped out, and Eglinton removed one slate, placed a scrap of pencil on the second slate, and then replaced, not the first, but a third slate lying close by on the table. Might some apparently trivial (to the conception of Mr. Massey) incident of this kind have been completely forgotten by Mr. Massey when he was writing his account, so that the preceding and succeeding events became joined in his remembrance? So far am I from thinking this impossible, that I regard it, owing to my experiences with other witnesses, as not even improbable. Indeed I think it even possible that Mr. Massey's holding of the slates might have been interrupted by some analogously "trivial" incident. Statements most express and definite, made by an honest witness, may, as I have already pointed out, be erroneous from other causes than simple mal-observation or the unwitting interpolation of "specific and positive acts of perception, " though they may be erroneous from these causes also. They are often due to other and more frequent forms of the universal weakness of human memory, and may be the result of transposition in the order of events, or of a mere and sheer lapse.

Concerning the possibilities, then, either of mal-observation or of lapse of memory, I traverse Mr. Massey's assertion that "the witness *could* not innocently use terms *expressly* and *definitely* inconsistent with what really happened," which I must characterise as an assumption completely destroyed by the reports of Mr. Davey's performances.

(c) But let us now admit, for the present purpose, that Mr. Massey's account of the incident is correct. He adds, in his report in *Light* (p. 159), that "as writing by the medium himself at the time is absolutely out of the question, there are only three other conceivable suggestions as opposed to occult agency." I do not of course question

Introduction

the veracity of the witnesses; and the other two suggestions, which Mr. Massey offers reasons for rejecting, are "a change of slate," and "concealed writing brought out by heat." Another suggestion has since, apparently, occurred to him.

> "As it is imaginable that a thin sheet of slate, already inscribed on one side, might be loosely fitted into the frame of one of the slates used, clean surface uppermost, so as to fall into the frame of the other slate, written side uppermost, when the first was placed upon the second, it is fortunate that I was able to exclude that suggestion by my possession of the slate on which the writing appeared, which, by-the-bye, was wrapped in paper, either by myself or by Mr. Eglinton—under my eyes, at my request, and carried away by me, *immediately* after we had examined the writing, the sitting being then closed." (*Proceedings*, Part X, pp. 88-9.)

There is clear indication that Mr. Massey had not contemplated this possibility at the time—two years earlier of the sitting, and it might be contended that the slate which Mr. Massey took away was not that one of the two slates upon which the writing was first found, but another similarly inscribed slate, which Eglinton had provided for the purpose; it probably often happens that sitters request permission to take away the slates upon which writing has appeared. Mr. Massey would apparently guard against this hypothesis of a *subsequent* change of slate by the last sentence of the passage quoted above, but no importance can be attributed to this, if only on the ground that Mr. Massey was unaware, at the time of the sitting, of the possibility to which he has more recently drawn attention. If the sentence in question expresses—not some record made by Mr. Massey on the day of the sitting, but—Mr. Massey's *remembrance after two years*, and if he seriously means it to exclude the hypothesis before us, there would be a vaster divergence than I have hitherto supposed between Mr. Massey and myself as to the ordinary psychology of memory; and I should, especially when I recall the rigorous signification with which Mr. Massey professedly uses the word *immediately*, class the sentence as an instance of *bona fide* transfiguration of the same character as some which I supposed in Mr. Davey's reports of Eglinton. (*Journal* of the S.P.R. for November, 1886.)

But I may point out that there is another suggestion still, which does not appear to me to be inconsistent with anything that Mr. Massey has said. The slate upon which the writing was found may have been an ordinary slate, and may have been taken away by Mr. Massey, and *the other slate*, though apparently ordinary, may have been a trick slate, with room for a false flap that could be fixed, if necessary, by a spring, and that was also adapted to fit the ordinary slate, and which had been placed so as to cover the prepared writing. The details of the trick will be obvious.

Mr. Massey's case is thus, in my opinion, vitiated on three separate grounds, by the considerations due to the possibilities of mal-observation, lapse of memory, and trick mechanism. I cannot, therefore, attach much value to his opinion, which, when his record appeared in *Light*, was:—

"I am as satisfied that this was a genuine phenomenon as I am that the words on this paper are of my own writing."

But while in this instance Mr. Massey's confidence in 1884 proves to have been misplaced, in consequence of his ignorance of a possible piece of trick apparatus, I do not think that his high estimate of the evidence for "psychography" generally is invalidated chiefly either by ignorance of this kind, or by his large trust in human observation. It is invalidated chiefly, I venture to think, by his *à priori* presumption that honest witnesses cannot use terms "*expressly and definitely* inconsistent with what really happened"; and he could not have made this presumption had he given due weight to the possibilities of memory illusion. I have already shown what small lapses of memory may have made his own specific and positive assertions erroneous; and it is obvious how, by simple omissions and transpositions, without any pure interpolations at all, the record of an honest witness may be rendered full of the most fundamental misstatements. Much of Mr. Massey's paper, as he himself says, was "an attempt to show" that "the supposition of such descriptions as he ' and others have given of Eglinton's slate — writing being given of the performances of an avowed conjurer is an impossible one." (*Proceedings*, Part X, p. 108.) The reader may decide for himself whether this "impossibility" has been realised or not. For my own part, I maintain that the reports which follow are a practical and complete rejoinder to the considerations alleged by Mr. Massey in his reply to the position advanced by Mrs. Sidgwick.

I may here again draw attention to some of the difficulties (which I pointed out in the *Journal* of the S.P.R. for October, 1886) in the way of obtaining adequate reports of "slate-writing" performances which are known independently to be the result of conjuring, and to the manner in which these difficulties have been partially, though not completely, overcome by Mr. Davey. In the first place, proficiency in the production of apparently "occult" slate-writing requires not only practice in the manipulation of slates, &c., but a lengthened experience of sitters, which cannot be acquired in a short time by a person who is chiefly occupied with other business. It is hardly to be expected that accounts of a novice's phenomena should compare for marvellousness with the results of "old mediumistic hands" like Eglinton and Slade. Still, Mr. Davey was able to devote much of his time during the latter part of last year to the improvement of his methods, and has thus been able to produce results which in quality, if not in quantity, may fitly be compared, for the purposes of our inquiry, with the productions of the best professional mediums.

In the second place, I must repeat that it is impossible to induce the same peculiarity of mental attitude in the sitters with a professed conjurer, as they would have assumed had they been sitting with a professional medium. I think I may safely say that not a single person of all those whose reports were published in the *Journal* for June felt certain beforehand that Eglinton's performances were explicable by conjuring; indeed, I may go further and say that nearly all, if not all, thought it not improbable that the phenomena were genuine, and that most of them had been strongly impressed by reports which they had previously heard or read. Now the evidence of a person holding this attitude is likely to be of decidedly less value *cæteris paribus* than that of a person who fully believes that he is watching a conjuring trick. I do not mean merely that there is a reluctance on his part to say or do anything which may imply a direct suspicion of the honesty of the "medium," or that, so far as his attention is directed at all, it is too exclusively occupied with the observation of the conditions at the time when the "occult" agency is *supposed* to be actually producing the writing; though from these causes also, in many cases, his testimony is likely to be less reliable. What I mean is that the idea of communication from the "spirit-world," or of some supernormal power in the "medium," will, in most persons, possess activity enough, even before any results are obtained, to interfere more or less with the observation of the conditions involved; and after the results are obtained, the dominance of the idea will frequently be great enough to contribute very materially to the naturally speedy oblivescence of many details of the sitting which were hardly noticed at the time of their occurrence, which in the course perhaps of an hour or two have dimmed out of recollection, but which, nevertheless, would have suggested the secret of the trick. Under this head I may also refer to the fact that the conversation held by the sitters with a professed conjurer will probably be of less avail in distracting their attention than if they were sitting with a "medium" with any the smallest expectation that "occult" phenomena might occur. In the former case they are well aware that the conversation is for the express purpose of distracting their attention from the movements of the conjurer; in the latter case, they endeavour to a certain extent to occupy the mind—according to instructions—with matters foreign to the sequence of events then and there transpiring.

In the third place, comparatively few persons are willing to write out reports of slate-writing experiences with a full account of the supposed test conditions, if they have any suspicion that the writing has been produced by mere conjuring. They are afraid of appearing ridiculous, and in this dread, if they are persuaded to write reports at all, they write them with a meagre allowance of detail, and with an abstention from dogmatic statement. No doubt the fear of ridicule has deterred many persons from writing reports on behalf of the professed "medium," but we must not disguise from ourselves the fact that when this fear has been overcome by the enthusiasm which often accompanies the formation of a new belief, the reports then are less to be trusted, by reason of that very enthusiasm. Analogous to that undeliberate warping of evidence which arises from the desire to justify the adoption of a new faith and to aid in proselytising others, is that which arises from the desire to strengthen the grounds of a conviction which has already been fully formed; the spiritualistic bias has been much more operative in transfiguring the

Introduction

accounts of mediumistic phenomena than most Spiritualists would be willing to admit. Possibly a wider experience may result in our finding a counterpart to this in the testimonials to professed conjuring performances, but my experience hitherto leads me to think that such a result is highly improbable.

These difficulties have been partly obviated by the fact that many of Mr. Davey's sitters were not informed until after they had written their reports that his phenomena were due to conjuring. It has, nevertheless, weakened some of the accounts, that Mr. Davey felt himself restrained from asserting that the phenomena were produced by "spirits"; he frequently asserted that they were *not* produced by spirits, but that he preferred to adopt the ordinary procedure of the spiritualistic medium; and, refusing to offer any explanation, thus shrouding their origin in convenient mystery, he usually requested his witnesses to determine the causes of the phenomena for themselves, the possibility of trickery included. Sometimes, indeed, he assured his sitters that he would take advantage of any carelessness which they displayed; and on one occasion he informed a sitter to whom the locked slate was afterwards entrusted and who placed it in the tail of his coat, that if he obtained an opportunity he would even take the slate from the pocket of this very sitter, and write upon it surreptitiously. Later in the sitting this witness, who I may say had been previously impressed by some writing obtained at a sitting with Eglinton, was called upon to produce the locked slate, and place it upon the table; yet so much had he been carried away by the appearance of the writing just produced by Mr. Davey on a single slate, that he had, for the time, completely forgotten the existence of the locked slate, and did not remember what he had done with it until the other sitters reminded him that he had placed it in his own pocket. This is a very extreme case of the influence of a mental attitude which has undoubtedly in some degree been very prevalent among sitters for mediumistic phenomena.

Not that it is by any means universal; and this leads me to say that one great advantage which some mediums, especially Eglinton, have used very freely, has been foregone by Mr. Davey. He has not been withheld from producing phenomena by the apparent observancy of the sitters. My own choice of sitters for Mr. Davey was determined chiefly by the desire to obtain educated and intelligent witnesses who did not know certainly that Mr. Davey's performances were conjuring, and who could be relied upon to write out a detailed account soon after the sitting.[17] Mr. Davey's unfortunate ill-health, soon after I first met him in September of last year, alone prevented him from obtaining a much larger number of reports. The witnesses were invariably urged to write their accounts as soon as possible after the sitting, while the occurrences were still fresh in their memory. It will be difficult for persons who have not had experience in making records of this kind to realise how quickly even incidents which have been recognised as important fade out of memory. And nothing betrays more fully the misappreciation of the value of the testimony we are considering than the reliance which has been so commonly placed upon detailed reports written weeks, months, and even years after the séance. I may refer here to the worthlessness of second-hand accounts and of abridged accounts. A careful original account, however faulty—if written shortly after the sitting, will often be found to contain some clues to the trick operations of the medium, which may be entirely absent from the later account. An instance of this was furnished by one of the reports printed in the *Journal* of the S.P.R. for June, 1886. The report was one sent to our Society by three of our Corresponding Members in St. Petersburg. It appears to have been condensed in a German magazine, *Neue Spiritualistische Blätter*, the "principal occurrences" only being mentioned; and a translation of this condensed account is given in *Light*, September 25, 1886. From this later version of the séance a series of incidents which indicate, as I think, how the chief trick was performed, are entirely omitted; and writing, which according to the original report is described as having been obtained on an ordinary slate, is described in the later version as having been obtained between sealed double slates.[18]

Ceteris paribus, the testimony of two witnesses to the same occur-

---

[17] I had another reason as well for requesting Mr. Davey to give a sitting to Mr. Padshah. I knew that Mr. Padshah had had some unconvincing sittings with Eglinton. Mr. and Mrs. Russell were unexpected witnesses; Mr. Legge was also an unexpected witness. A séance was given to Mr. Dodds because he had scoffed mercilessly at Mr. Legge's account of his experience.—See SITTINGS II, III, IV, VIII.

[18] See Supplement to the *Journal* for December, 1886, p. 516.

rence is rightly regarded as better than the testimony of one. We must, nevertheless, be cautious in the application of this principle to the testimony for "psychographic" phenomena. A reference to the reports which follow will show that wherever separate accounts are given of the same sitting, the witnesses are never in complete agreement, and they sometimes differ on important points. This we found also to be the case where independent accounts were given of the same sitting with Eglinton (*Journal* for October and November, 1886). In the large number of cases, therefore, where the detailed record of a sitting has been made by one only of two or more witnesses, and the record thus made has been every word of it endorsed by the other witness or witnesses, we have another proof of that plasticity of memory to which I have drawn attention. It must be concluded that in the majority, if not all, of these cases, some of the *remembrances*, whether true or false, of the later co-signatories have become transformed to fit the remembrances of the recorder, and this, perhaps, in most instances, by the mere reading of the record. It is of course not easy to obtain accounts which shall be absolutely independent, since this requires that no communication of any kind connected with the phenomena should pass between the witnesses after the commencement of the séance until the reports have been written, and probably only an approximation to this absolute independence has been attained where separate accounts have been given of the same sitting with Mr. Davey.

It may occur to some ingenious readers of the reports which follow and the notes given in the Appendix, that owing to illusions of perception and memory, Mr. Davey and myself have given misdescriptions of what really happened at the séances, that the other witnesses are right and we are wrong.[19] It is unnecessary to explain the "psychical" conditions under which Mr. Davey and myself were sitting, so different from those of the other sitters; I may refer such readers to the opinion of persons versed in conjuring tricks, and familiar with the misdescriptions given of them by the uninitiated—and especially to the opinion of Mr. Angelo J. Lewis (p. 189). I repeat that the "psychographic" phenomena described in the following records are conjuring, and only conjuring, performances, and I may add that I was very

---

[19] Mr. Davey agreed with my notes except where I have stated to the contrary.

careful myself, in the séances where I was present, to take the part of an ordinary sitter, and to avoid doing anything which would assist Mr. Davey in the smallest degree. Indeed, Mr. Davey would have preferred my absence, as part of the task which I had set myself was to watch Mr. Davey's movements at the critical moments, in order that I might give my independent testimony concerning the mode of production of the phenomena.

These phenomena, as described, may well seem marvellous enough to demand the hypothesis of occult agency: writing between conjurer's own slates in a way quite inexplicable to the conjurer,—writing upon slates locked and carefully guarded by the witnesses,—writing upon single slates held by the witnesses firmly against the under-surface of the table,—writing upon slates held by the witnesses above the table, answers to questions written secretly in locked slates,—correct quotations appearing on guarded slates from books chosen by the witnesses at random, and sometimes mentally, the books not touched by the "medium," writing in different colours, mentally chosen by the witnesses, covering the whole side of one of their own slates, messages in languages unknown to the "medium," including a message in German for which only a mental request had been made, and a letter in Japanese in a double-slate locked and sealed by the witness, the date of a coin placed by the witness in a sealed envelope correctly written in a locked slate upon the table, the envelope remaining intact, a word written between slates screwed together and also corded and sealed together, the word being chosen by the witness after the slates were fastened by himself, &c., &c. And yet, though "autographic" fragments of pencil were "heard" weaving mysterious messages between and under and over slates, and fragments of chalk were seen moving about under a tumbler placed above the table in full view,—none of the sitters witnessed that best phenomenon, *Mr. Davey writing.*

A few words remain to be said as to the principle which has been followed in adding or withholding notes explanatory of the *modi operandi* adopted by Mr. Davey.

The object of the notes given in the Appendix is not to explain the tricks, though explanations of some of the incidents are given or suggested. To explain the tricks would in itself be of little advantage

Introduction

to the investigator of the "physical phenomena" of mediums, since many methods of producing "psychography" may exist besides those which Mr. Davey has employed; and were all of those in present use to be made public property, others would doubtless be invented, and accidental opportunities for producing successful illusions would still arise. I may point out moreover, —and this is a consideration frequently overlooked—that it is a great mistake to suppose that specific verbal explanations, in as far as these are possible, of all the methods practised by "mediums," would effectually check deception even by those very methods; there are tricks that can be explained to a witness, and then performed in his presence without his detecting them, and while he imagines himself to be fully on his guard against them.[20] It would be a still greater mistake to suppose that explanations of the methods in use would convince those who have testified from personal experience to the genuineness of the "psychography" of Eglinton, Slade, &c., that such methods were used for the production of the phenomena which they witnessed. They will scarcely be likely to *remember* the occurrence of events which they perhaps never observed at all, or observed only partially and erroneously; which, whether correctly or incorrectly observed, they have afterwards continually misdescribed or completely forgotten; and which, in many cases, would be distinctly excluded by the acceptance of their testimony as it stands. Further, it must be said,—and this will become obvious to the careful student of the reports that no description of particular movements or peculiar apparatus would suffice as an exhaustive explanation of the performances recorded; the best part of the trickery is in truth indescribable, it is as fluent and uncertain as the shifting attention of the witnesses, and varies with the variations of their temporarily dominant expectations and emotions.

The object of the notes, then, is to show to investigators the kind and degree of mistakes which may be made by educated and intelligent witnesses in recording their impression of a performance the main lines of which are planned with the deliberate intention of deceiving them, but few, if any, of the details of which can be described as absolutely fixed.

---

[20] I have actually seen this done.

The notes, as given, might suggest to the ordinary reader some unfounded conclusions. They seem in some cases to indicate so much carelessness on the part of the sitters, to open out such easy possibilities of trick for the conjurer, that the reader may fancy that the witnesses were unusually gullible, and that Mr. Davey had in reality little to do. To this the supporters of Eglinton may probably add that an "experienced Spiritualist" would have run no risk of being similarly deceived. These conclusions, I have no doubt, would be mistaken. In the first place, the witnesses would certainly not have been objected to before actual trial as of less than average competence; nor do I think that as a matter of fact they have shown less than average acumen or care. They may be taken, too, as a fairly representative group, including, as they do, successful men of business, men of ordinary university training, electrical engineers, members of the legal and educational professions, &c.; they include one professional conjurer, and others as Mr. Padshah and Miss Symons—who had given some previous attention, as their accounts may sufficiently show, to the risks of mal-observation on such occasions as these. Of the great treachery of memory, indeed, the majority of Mr. Davey's sitters have been unaware, but of this the witnesses of other "slate-writing" performances have, unquestionably, been at least equally unaware. In support of this it is enough again to remind the reader that in all the cases where separate accounts, more or less independent, of the same sitting were given by the witnesses of Eglinton's séances recorded in the *Journal* of the Society for Psychical Research for June, 1886, a comparison of the accounts elicits the fact that the witnesses exhibited forms of memory—illusion precisely parallel to some of those which have been exhibited by the witnesses of Mr. Davey's performances.[21]

And be it observed that mistakes of the kind illustrated in the notes are not, prior to special study, any proof of obtuseness or particular deficiency of memory. Mr. Davey himself, before he had studied the "slate-writing" forms of conjuring, was a decided offender in this respect (see *Journal* of the S.P.R. for October and November, 1886); and I have pointed out my own shortcomings and those of my colleagues with, I hope, an unsparing hand (see Supplement to the De-

---

[21] See the criticism of the evidence in the *Journal* for October and November, 1886.

cember number of the *Journal*). I do not, in point of fact, think that this unpreparedness and in observancy of mind, in the presence of a conjurer, is a thing of which anyone who is not familiar with the tricks already need be ashamed. I have a strong suspicion—founded by this time on a pretty wide experience that those of my readers who may be most disposed to deride the simplicity of the witnesses now to be cited, would themselves have come out of the ordeal no better than Mr. Rait, and not so well as Mr. Padshah.

But, although mankind—as I believe—are thus inattentive and careless and forgetful beyond their own common notion of themselves, it must not be supposed that it is an easy task to play upon their inattention and forgetfulness as Mr. Davey has done. On the contrary—without making any claim for him of inimitable skill—it must be pointed out that his achievements, simple as they may seem, and by reason of their very simplicity, belong not to a low but to a high class of conjuring performance. For the psychologist, and quite apart from the question of effect producible on large audiences, there may be said to be three classes of conjuring of progressively deeper interest. Least interesting to the psychologist are the effects which depend on machinery or apparatus, great as the mechanical skill involved may be. More interesting are those which depend on pure prestidigitation, on an assured competence to perform some given action so swiftly and cunningly that the spectator will fail to see it. But higher still in psychological interest comes a class of conjuring performances which consist not so much in eluding the perceptions of the witnesses by the speed and dexterity of one's own movements, as in gently inducing them, by means different in each case, to bewilder and entrap themselves. The conjurer here wins as by the adroitness of a clever thief, who poses for then once as a detective, persuasively points out the most subtle and efficient precautions against robbery, and all the while is emptying your pockets. In this last class of performance, the better the conjuring is, the less of it is needed; and its greatest triumph is when the spectator's mind has been brought into such a state that he—so to say—does his conjuring for himself, and stands astounded at his own interpretation of some entirely obvious phenomenon. Thus Mr. Padshah, in one of the most instructive of the incidents below to be recorded, turned the conjurer's very scrawl into a crowning success, and read *Books* into *Boorzu*.

Once more; if it be claimed in any quarter that "experienced Spiritualists" would have been able to detect Mr. Davey's methods more easily than the witnesses actually adduced, I trust that I may be allowed to say, without giving offence, that to my mind the presumption is strongly the other way. I have already explained that the power of detecting conjuring tricks is not a test of general capacity, but rather a result of having studied similar conjuring tricks beforehand. The experience "involved in much séance—going is assuredly not an experience that makes in this direction; the mental attitude induced is the very worst possible for the discovery of trickery, and it has been on "experienced Spiritualists" that those mediums have thriven whom mere ignorant outsiders have afterwards caught in palpable fraud. I do not, of course, mean that no "experienced Spiritualist" has ever caught a rogue out; but such exhibitions, for instance, as those of Haxby and Firman (see *Proceedings*, Part X, pp. 60-62), certainly show that it is not to the circles of devout believers that we are to look for detection of even the grossest and most transparent imposture. But I am anxious not to be supposed to assert that" experienced Spiritualists " are less acute than other men. I must yet again. emphasise the fact that the difference depends on previous attitude of mind;and there is plenty of evidence to show how many minds at how many séances, have been in much the same condition as Mr. Padshah's when he recognised the convincing word *Boorzu*. I will cite one instance only, which exemplifies the influence of the Spiritualistic bias even beyond the sphere of professed mediums. is a letter written by perhaps the most experienced of all Spiritualists,endorsing an opinion held by a man than whom none more eminent, none more widely or justly respected, has ever avowed adherence to a belief in Spiritualism. "M.A. (Oxon.)" wrote as follows to *The Medium and Daybreak* of August 24th, 1877:—

I am glad to see that Mr. Alfred Wallace agrees, after seeing Lynn's medium,with the substance of my letter in your issue of July 6th. Given mediumship and shamelessness enough so to prostitute it, and conjuring can, no doubt, be made sufficiently bewildering. It is sheer nonsense to treat such performances as Maskelyne's, Lynn's, and some that have

Introduction

been shown at the Crystal Palace, as "common conjuring." Mr. Wallace positively says, "If you think it is all juggling, point out exactly where the difference lies between it and mediumistic phenomena. " (See *Proceedings*, Part X, p. 66 and *note*.)

Few readers indeed will question the proved sagacity, the absolute straightforwardness, of the illustrious naturalist whose statement is quoted by "M. A. (Oxon.)" Still fewer perhaps will think that, without a strong mental predisposition, the author of *Essays on Natural Selection* would have committed himself to the view that unembodied spirits ran Dr. Lynn's entertainment at the Westminster Aquarium. For myself, I can but repeat his challenge in another sense, and say— let the experienced Spiritualist "point out exactly where the difference lies between 'Mr. Davey's performances' and mediumistic phenomena."

EXPERIMENTAL INVESTIGATION.

By S. J. Davey.

For some time past I have practised "slate-writing," and have given up much leisure time to the subject, with a view to discovering how far ordinary witnesses can be deceived by conjuring performances. I have received reports of my experiments from various persons, and these I subjoin, with comments, in some cases, as to lapses of observation or memory on the part of the witnesses. Lest there should be any misunderstanding I must explain what induced me to take up the subject, and the general conclusions to which I have been led.

Readers of the *Journal* of the S.P.R. are aware that I sent reports of sittings with Eglinton to our Society in 1884, and that I had previously sent reports of the same sittings to the periodical Light. I do not now attribute any value to these reports as proving the reality of so-called "psychography," for reasons which will appear in the sequel.

My chief interest in Spiritualism generally was awakened by an experience of my own, which was as follows:—In 1883, owing to a serious lung complaint, I spent several months at a Continental health resort. During this visit, one of my companions died under circumstances of an unusually distressing character; and another friend and myself had been in frequent attendance upon him during his last illness. His body was subsequently dissected, in the presence of my other companion, Mr. C. Three weeks after this I was startled one night by seeing what appeared to be the face and form of my deceased friend under circumstances that greatly surprised me, and the next day, whilst visiting Mr. C., who lodged in the same hotel as myself, he informed me that he had that night experienced a remarkably vivid dream in which he had seen our deceased friend. I then for the first time related to Mr. C. what had happened to myself.[22]

On my return to England I began to devote some attention to the study of alleged psychical phenomena, and I perused several works. relating to the subject, including Zöllner's *Transcendental Physics*, *Psychic Force*, by Professor Crookes, *Miracles and Modern Spiritualism*, by Alfred Russel Wallace, *The Debatable Land*, by Robert Dale Owen, *Psychography*, *The Report of the Dialectical Society*, &c., and I formed a circle of friends for the investigation of the alleged phenomena. During my first experiments I found myself affected a good deal by involuntary movements which I could not then account for, though I now have little doubt they were caused simply by nervous excitement; however, nothing of any significance happened, and it was at this stage of my investigation that I made the acquaintance of Eglinton, of whose so-called "psychography" I had heard. At the conclusion of my first séance with Eglinton, which took place in June, 1884, I could not account for the phenomena except on the Spiritualistic hypothesis, and I was led to believe, from the "communications" which I then received, that I possessed psychic powers. My second séance with Eglinton, on October 8th, 1884, was a failure, but my third, on October 9th, 1884, was a success. I was somewhat excited at these results, and even contemplated making a collection of cases to convince the unbelieving world. On October 9th, 1884, the supposed invisibles informed me that I had "developed my own powers to an appreciable extent, owing to their former advice." Now, between my first and third séances I certainly had experienced privately one or two incidents which I then regarded as genuine psychical phenomena, and I will briefly relate one of these experiences.

One afternoon in September, 1884, I took two slates and determined to experiment alone. I held them together with a small pencil grain between. I was in my library; the slates were taken out of a private box by myself; I glanced at them and placed them in the position above described. In the course of some few minutes I lifted up the slates and examined them, and found the word "Beware" written in

---

[22] I have since had some correspondence with Mr. C., who does not look upon the incident as anything more than a dream coincidence. At the time, I attached particular significance to my own experience, as my friend, when alive, had discussed the question of Spiritualism with me.

large characters across the under side of the upper slate. My astonishment at this cannot well be described, as I felt convinced I had previously thoroughly examined the slates, and I took the first train to London, and showed them to my friend Mr. X. (see *Journal* for October, pp. 435, 436). He agreed with me in saying it was almost incredible. I then attributed the above, and one or two kindred phenomena, to the action of an abnormal power proceeding from myself.

It was afterwards proved to me that these experiences were neither more nor less than simple hoaxes, perpetrated by some of my friends. In the case of the particular incident which I have described, the slate had been tampered with during my previous absence from home. I have no doubt that, not suspecting any interference with my slates, I had not thoroughly examined them immediately before sitting, as I supposed myself to have done. Another incident of a somewhat ludicrous character may be mentioned here. I had bought a trick slate, which had been sold to me as an explanation of the process used by mediums. I thought, however, that this was scarcely true, as the trick seemed to be a very palpable one. I had put this slate away in my drawer with the other slates containing the writing of Eglinton's supposed spirits. One morning, on going to this drawer, which I usually kept locked, I found the following words, or something to the same effect, written across the false surface of the trick slate: "We object to your learning trickery." I then compared this writing with some on Eglinton's slates, and found it apparently identical. I was naturally somewhat amazed, and I did not then for a moment suspect that my friends were hoaxing me, and that the above sentence had been written in careful imitation of the writing on Eglinton's slates. Also, during séances held privately, I continued to be frequently seized by spasmodic movements when I believed "uncanny" manifestations were about to take place. As a conjurer, I have been since amused sometimes at similar convulsions in others during my conjuring performances, when the sitters have supposed that the writing was being produced by supernatural means; my own shudderings during these performances being, of course, part of the trick.

I had several other séances with Eglinton after October 9th, 1884, all of which proved blanks, except one held on January 15th, 1885. One of my friends who accompanied me to this sitting assured me he had actually seen Eglinton imitating the sound of writing at the time when I thought a long communication was being written. I endeavoured to be more watchful at the two sittings which I had after this, the final séance being on June 25th, 1885; but at neither of these did any results occur, although I did not inform Eglinton of the information I had received. However, partly in consequence of my friend's conviction that Eglinton's performances were only tricks, I began, after getting no further results, to apply myself anew to see what could be produced by conjuring. I then met with an individual who professed to sell me "secrets," which he gave me to understand he had procured from an American medium. I also bought one by which words, &c., could be made to appear on the flesh after it was rubbed over with burnt paper. This trick has evidently been exhibited by Eglinton as a "mediumistic phenomenon" (See *'Twixt Two Worlds*, pp. 52, 54). I soon made use of the knowledge thus acquired by performing before friends and acquaintances, and I found that even at that early stage of my practice many of them could be deceived as to my real *modus operandi*. Eglinton has attempted to give particular validity to the accounts of my successful séances with him in 1884, claiming my testimony as that of one who had "specially studied and practised the art of simulating the slate-writing phenomena under conjurers' conditions" (*Light*, July 31st, 1886). I have already pointed out elsewhere that I was not an expert in 1884, when I wrote the reports in question, which Eglinton describes as "among the most favourable and decisive which have appeared." The extent of my knowledge on this subject at that time will be found described by myself in *Light*, August 21st, 1886, as follows:—

I went to Mr. Eglinton on June 30th, 1884, and I do not remember ever having previously performed a single conjuring trick as applied to slate-writing, and also the question of conjuring in any other form had in no way interested me. Previously to my second séance, October 9th, 1884, I made some three or four attempts with a thimble, pencil, and a slate held under the table, and with a trick slate made of cardboard, with a movable flap and blotting-paper.

I noticed that many persons made statements concerning my performances, as to the conditions of the production of the writing, which were just as emphatic as I made in my own reports about Eglinton, and I also noticed that nearly all these statements were entirely wrong.

Even when I sometimes revealed the fact that I was merely a conjurer, the reply which I frequently received was something of this kind: "Yes, you may say it is conjuring, but it could not have been done by that means when I did so-and-so"(describing a supposed test) "and yet we got the writing all the same." The following extract from a letter to me is typical of the views taken by several of my investigators:—

I certainly think your slate-writing quite equal to what we saw at Nottingham-place [with Eglinton], but till I see how it is done, and it is thoroughly explained to me, I cannot give up my belief that you yourself employ more than sleight of hand for your results. You see I am a St. Thomas.

As I went on I was gradually forced to the conviction that my own reports about Eglinton were just as unreliable as these statements about myself, although I was not then aware of the serious discrepancies between them which Mr. Hodgson has lately pointed out in the *Journal* (October and November, 1886).[23] In consequence of the change which was taking place in my opinion, I wrote, on July 30th, 1885, to Mr. Farmer, requesting him for "private reasons," not to make any reference to myself, either directly or indirectly, in the work about

---

[23] A critic, "C. C. M.," says in *Light,* January 22nd, 1887:—"Of course in very many cases the modus operandi could and would be explained, at least privately to the witnesses, to their complete satisfaction. These would be really trick cases, as to which the antecedent reports will always have some evidential defect, discoverable by a careful critique without any presumption at variance with the distinct and definite statements of the witnesses (except in the case of witnesses whose veracity, or capacity for ordinary observation, is questionable), and such, therefore, as would not be adduced in any judicious selection of evidence to prove the genuine phenomena." Now, in a previous number of *Light* (August 14th, 1886) Mr. C. C. Massey asserted, concerning some of my testimony to Eglinton's performances, that "there is no room for the hypothesis of innocent misdescription, which might afterwards come to be recognised as such by the witness himself." I believe, nevertheless, that the performances of Eglinton, which I endeavoured to record, were due to trickery on his part, and that my reports, adduced by Mr. Massey in his "judicious selection of evidence," were vitiated by " innocent misdescription. " I think it not at all unlikely that I was guilty of the amount of misdescription hypothetically attributed to me by Mr. Hodgson, because precisely similar misdescriptions have been given of my own performances by witnesses whose intelligence and acumen are certainly not inferior to mine, and whose veracity is unquestionable.

Experimental Investigation

Eglinton (*'Twixt Two Worlds*), which he was then preparing for the press.

From a study of various exposures of slate-writing mediums, and other incidents which have been privately brought to my notice, I cannot now entertain a doubt that they have frequently practised deception; and whether it is a fact that they, nevertheless, occasionally obtain the help of "spiritual" beings, or manifest supernormal powers, is a question upon which I have good reasons for being now very sceptical, though I do not of course profess to knowhow all the slate-writing tricks are performed. Indeed, only last month (February, 1887), I was informed of a special *modus operandi* employed by an American medium, Mrs. Simpson; this method was entirely unknown to me,nor do I think I should have discovered it myself.

Until recently I had not endeavoured to obtain written reports from persons who sat with me, and I was desirous of obtaining them under as nearly as possible the same conditions, as regards the mental attitude of the sitters, as those obtained by professional mediums for slate-writing; I did not wish people to know with absolute certainty by my own professions beforehand, that the slate-writing was only conjuring, though I urged them to treat me as a conjurer, to use tests, and take precautions against trickery, &c. Few persons would imagine how difficult it is for ordinary witnesses to accurately record a "slate-writing" séance, even if they are very careful and quick observers; and how prone the majority of witnesses are to exaggerate or distort records of events which they believe to be of an abnormal character. In consequence of the prominence given in certain quarters to my name in connection with "slate-writing," I assumed the professional name of David Clifford.[24] The desirability of this step may be illustrated by the following incident: A short time ago, at a séance, I met a gentleman who spoke in very disparaging tones of the performances of a certain amateur conjurer known as Mr. A., and who remarked to the effect that the statements of Mrs. Sidgwick (*Proceedings*, Part X, pp. 67-70) as to this conjurer's powers did not in the least explain the subject of "psychography." At the conclusion of my performance this same gentleman(who knew me only under the name of Clifford) declared in

---

[24] By the use of this term I do not mean to imply that I have ever demanded any fee for a séance; I have never accepted any recompense whatsoever.

my presence, and in that of his co-investigators, that the experiments he had just witnessed were more conclusive as to the existence of supernormal phenomena than those he had witnessed in the presence of a well-known professional medium. Had he then known I was Mr. A., the "amateur conjurer," I do not think he would have shown such enthusiasm as regards the "incomparable" nature of my phenomena.

In some of the reports which I received I was described as Mr. A., and in several others as Mr. Clifford; in these cases I have substituted my real name, for the sake of clearness.

I think it would be no easy task to expose an expert in slate-writing, provided he had made up his mind not to give his investigators the chance of doing so. A practised conjurer in this particular branch of his profession soon acquires a sufficiently keen insight into character to know when there is no risk of detection. If the performer has any reason to think that any part of his trick will be seen, he can take refuge in a blank séance; nor would it generally be the case that if the trick were partly performed the observance of strict conditions by the sitter would result not merely in failure, but in exposure, as Mr. Massey seems to suggest. (*Proceedings*, Part X, pp. 93, 94.) I have, several times, had to deal with this danger, and have always been successful.[25] Of course, cases will arise when, if the right steps are taken by the sitter, exposure will result; and this is precisely what has happened on more than one occasion, with, for example, Dr. Slade. There is one danger to which I think a conjurer is liable, unless he is very careful, viz., to give too little credit to the shrewdness of a sitter, just as he probably often gives too much. The remedy obviously would be to increase the number of entirely blank séances. If I were forced to give blank séances to persons of whose keenness I was afraid, I should, of course, frequently give blank séances to others whom I had no reason to fear, and with whom I could produce marvellous phenomena whenever I liked. I have found, moreover, that a blank sitting occasionally, with an investigator who at other times gets good results, makes the phenomena look more mysterious than ever, and forms an additional reason in his mind for not attributing the phenomena to conjuring. A plan, I un-

---

[25] Since the above was written, an incident has occurred which some of my readers will probably think an exception; it happened at a sitting with Mr. Dodds, and I will refer to it in my comments upon his report.

derstand, that is very frequently adopted by a well-known American medium, is to simulate sometimes, in a very marked manner, the appearance of trickery in his slate-writing. Not unfrequently one of his investigators falls into the trap, observes what he supposes is a clear case of deception, and demands an instant exposure of the slate. The medium then protests against the "unwarrantable suspicion," and finally reveals the slate, to the chagrin of his would-be exposer, who of course finds it perfectly clean. Then, by a subtle process, the medium does write on the slate, to the subsequent amazement of his witness. From the account of a recent exposure by a lady Spiritualist in America, who detected Slade in the very act of writing, I understand that the speed with which he wrote on a slate held under the table greatly astonished the observer. I have good authority for believing that the account is to be relied upon. (See New York *Sunday Times*, July 5th, 1885.)

I may now briefly refer to the argument that "psychography" must be of an abnormal (or supernormal) character, since conjurers have been unable to explain the phenomena. My own opinion, as that of an amateur conjurer, has been claimed in its favour, but I have already pointed out that this is only a misrepresentation of the facts of the case, and that I was a deficient observer, and an ignoramus as regards conjuring, when I wrote the reports favourable to Eglinton. At the same time, I understand that certain conjurers have professed their inability to explain the slate-writing of some mediums by conjuring. But, after my own experiences, I am not at all surprised at this. That the testimony of a specially skilled conjurer *in this particular branch* is of value I do not deny, yet at the same time it does not, I think, follow that he must therefore know all the secrets, such as one with more experience might have acquired. If he is very confident of his own ability to find out any trick and cannot explain the *modus operandi* of the medium, he may possibly think it inexplicable by conjuring; and the remarks made by Mrs. Sidgwick at the close of her article in the *Journal* of the S.P.R. for December are particularly suitable to a case of this kind. A very good instance of this has come under my notice.

When Eglinton was in Calcutta, Mr. Harry Kellar, a professional conjurer, requested the "opportunity of participating in a séance, with a view of giving an unbiassed opinion as to whether," in his "capacity of a professional prestidigitateur," he could "give a natural explanation of effects said to be produced by spiritual aid." Eglinton eventually met Mr. Kellar, and the result was that Mr. Kellar came away utterly unable to explain by any natural means the phenomena that he witnessed; and he said that the writing on the slate, "if my senses are to be relied on, was in no way the result of trickery or sleight of hand." This occurred early in 1882, and Mr. Kellar's opinion still continues to be quoted in favour of the genuineness of Eglinton's phenomena. Yet I am not aware that Mr. Kellar, before sitting with Eglinton, had any special knowledge of the different methods of producing slate-writing by conjuring, and I have little doubt, after reading his account of a sitting in 1882, quoted in *Light*, October 16th, 1886, p. 481, that he was ignorant of at least some of these methods. But this does not seem to be my own view only; it seems to be that of Mr. Kellar himself, who since then has apparently turned his attention to slate-writing, and has changed his former opinion about the genuineness of the phenomena; he now professes to be able to "duplicate any performance given by mediums of whatever nature after he has seen it done three times." This was mentioned to me by an American gentleman whom I met recently, but I have also seen a notice of it in *Light* for March 28th, 1885, p. 147, from which I have taken the above extract; yet Mr. Kellar's former opinion, given, as I presume, when he was not a special expert in slate-writing, is continually quoted by Spiritualists, just as my own opinion, given when I was absolutely incompetent and knew next to nothing about conjuring in any form whatever, has been quoted as the opinion of a specially qualified conjurer.

I do not myself place much value upon the opinion of conjurers who have not previously become thoroughly versed in the ways of deceiving sitters in slate-writing; not only because of this incident in which perhaps Mr. Kellar's over-confidence in his own powers of detection led him into a mistake, although he has after long experience publicly proclaimed his disbelief in "mediumistic" phenomena, but also because I have myself been able to deceive a gentleman accomplished in general conjuring.

On August 26th, 1886, I received a letter from a well-known professional conjurer, whose programme includes several exposés of alleged spiritualistic frauds. In his letter to me this gentleman informed me that he had heard a great deal about my slate-writing, and was most anxious to witness the phenomena, as he had had séances with a well-known professional medium; and he politely requested an interview with me.[26] He was a stranger to me personally, but I at once offered to give him a séance, which was arranged for September 13th, 1886. At the conclusion of the séance he gave me his testimony as follows:[27]

September 13th, 1886.

I can see no explanation by trickery of the experiments in slate-writing I have seen performed by Mr. Davey this evening.

(Signed) _____

Some days afterwards he wrote to me as follows:—

September 24th, 1886.

It gives me much pleasure to add my testimony to that of many others you have, and I certainly can state that in some mysterious manner which to me seemed quite inexplicable, writing appeared on slates which I had purchased myself, which had been previously thoroughly washed, and while they were held together apparently very tightly. And it was specially remarkable that the writing was in the very colour I asked for.

(Signed) _____

---

[26] I have not here disclosed his name, as, since I have informed him of my conjuring powers, he has desired me not to do so. The names and addresses of all the writers of statements and reports are in the hands of the Hon. Secretary of the S.P.R.

[27] I had a curious experience with this gentleman. I asked him to think of a number. A number which I thought would be right was then, without his knowledge, marked on the slate by my process. I then asked him to tell me the number he had thought of. He said 98. I lifted up the slate and showed him the figures 98 that had been written before he had spoken. This may of course have been merely an odd coincidence, but the fact that I have had several somewhat similar experiences with other investigators led me to think that there might be something of the nature of thought-reading in it. I endeavoured to arrange some further experiments with Mr. ____, but his many engagements, and afterwards my serious illness, prevented our meeting again.

Another professional conjurer was shown my locked slate by an investigator, the writing having been allowed to remain, and on hearing the account of the witness, he offered an explanation, which was, however, entirely wrong; I instance his opinion merely for the sake of pointing out that his great knowledge of conjuring in general did not enable him to suggest an explanation which would I think have occurred to him if he had been skilled in the various special methods that may be used by conjurers in connection with slate-writing.

It has sometimes happened that an investigator, who knew beforehand that my performances were conjuring, has thought he had obtained a clue to my methods, but in nearly every case where I have suspected this, I think his discovery has only tended to perplex him more than ever. Whilst visiting Professor Henry Sidgwick at Cambridge some few months ago, I gave both Professor Sidgwick and Mrs. Sidgwick two séances for slate-writing. Amongst other phenomena, I obtained an answer on my locked slate, written underneath the question Professor Sidgwick had written. I had requested Professor Sidgwick to keep special charge of the slate. He afterwards concluded I had obtained some means of opening and writing on it, and he informed me as to when and how he thought I had done this. It is interesting to note that I did not in any way perform the trick in the manner Professor Sidgwick surmised, as I have since proved to him; he has informed me that my explanation was "completely unexpected," and he says:—

I was so satisfied with my own conjecture (difficult as it was for me to imagine it actually realised) that the method you actually used never occurred to me—nor anything at all like it.

To those of my readers who are specially interested in the subject, I may recommend a book entitled *The Bottom Facts of Spiritualism*, by Mr. John W. Truesdell, who seems to have had considerable experience in slate-writing.[28] He gives an interesting account in Chapter XVI. of a slate-writing séance recorded by Mr. L. W. Chase, of Cleveland, Ohio, and I have no doubt, after my own experiences as a producer of slate-writing, that Mr. Truesdell's subsequent version of the matter

---

[28] Published by Carleton and Co., New York.

is the true one. In the *Daily Courier* of Syracuse, New York, December 7th, 1872, Mr. L. W. Chase made the following statements:—

The medium (Mr. John W. Truesdell) then took up a common slate, and, after carefully washing off either side, placed it flat upon the table, with a bit of pencil, about the size of a pea, underneath. We then joined hands, and after the lapse of about ten minutes, under the full glare of gas-light, we could distinctly see the slate undulate, and hear the communication that was being written, a copy of which I herewith append: —My dear Brother,—You strive in vain to unlock the hidden mysteries of the future. No mortal has faculties to comprehend infinity.—CHARLOTTE.

The above lines were not only characteristic of my beloved sister while in the form, but the handwriting so closely resembled hers that, to my mind, there cannot be a shadow of doubt as to its identity.

In reference to a further event, Mr. L. W. Chase adds:—

A short communication from my mother (and in her own handwriting) was found plainly written.

I have quoted the above extracts since they serve to show how a person may be deceived in the matter of spirit identity; for Mr. John W. Truesdell, at the close of Chapter XVI., frankly informs his readers that he himself wrote the messages, and describes the methods he employed. The resemblance between the handwritings was, I presume, imaginary.

The fact that "messages" occasionally contain private family details, &c., is often quoted as a proof of the Spiritualistic theory in connection with slate-writing, but many persons would be surprised to find how frequently a slate-writing conjurer may become possessed of apparently private matters in connection with his investigators, and they should also not forget that peculiar chance coincidences sometimes occur. It is not very long since I met a gentleman who was a perfect stranger to me personally, and I depicted scenes to him that I knew had taken place many years ago, with an accuracy that utterly bewildered him, and I went into such private details of his family matters as convinced him I had a strange insight into his past life. Yet this was merely due to a chance coincidence. Some months previously these and other details had been incidentally mentioned to me by a person well acquainted with his history, and although he was not a public character, his name, in connection with the events of which I had heard, became somehow fixed in my memory. Nor is this the only experience I have had of a somewhat similar nature.

Then it must be borne in mind that when witnesses become deeply impressed with the wonder of the performance, they not unfrequently give way to a little natural excitement, and whilst they have laboured under the excitement I have picked up items of information from the witnesses themselves, which when reproduced by me at future séances have been declared "wonderful tests."

During the past few months I have given séances to many total strangers who have applied to me for sittings. In some cases I have given these performances away from my own residence, and I have requested the investigators to use all possible caution to guard against any trickery, leaving them, however, to make their own suppositions concerning the mode of production of the phenomena. Latterly I have stipulated that the sitters should write out reports as soon as possible afterwards; and upon receiving these reports I have informed them without delay that the phenomena were only conjuring. Formerly in some cases I had given the sittings over and over again to the same persons, with an occasional blank to stimulate their curiosity; nevertheless they never detected the *modus operandi*.

I shall now give nearly all the accounts that I have received, but before doing so I wish my readers to be clearly aware that the writing performances described in the following records were due to my own unaided powers as a "slate-writing" conjurer.[29] Two years ago I should have questioned the power of a conjurer to produce such records

---

[29] The only reports which I have not quoted are two by Mrs. Sidgwick, one by Mr. Hodgson, and three others. These last three resembled the majority of those which I have quoted; two of the writers desired me not to print their reports at all; the third desired me not to print his report unless it was a correct account of what occurred at the sitting. Mrs. Sidgwick was not only aware that I was a conjurer, but I had told her a good deal about my tricks before my first séance with her. Mr. Hodgson was also aware that I was a conjurer, and had some knowledge beforehand of my modus operandi; his report was written chiefly for comparison with an account promised by the gentleman to whom I have referred on pp. 85-6 as having exhibited such enthusiasm, but who, unfortunately, notwithstanding my repeated requests, has never sent me any report.

from ordinary witnesses as those which I now append, and that others shared my doubt in this respect is, I think, apparent from the following Editorial Note in *Light*, September 4th, 1886.

If he [Mr. Davey], or any other conjurer, can produce the *appearance* of the conditions which he seemed to observe with Mr. Eglinton, and the writing under such *apparent* conditions, so as to induce an inexperienced witness to write such a report as those he wrote himself, it will be time enough to talk of mal-observation as a possible explanation.

I shall begin by quoting a few brief statements of a general kind merely in illustration of the impressions left upon some of those from whom I did not exact a detailed report.

Statement of Mr. A. Podmore.

July, 1886.

A few weeks ago Mr. D. gave me a séance, and to the best of my recollection the following was the result. Mr. D. gave me an ordinary school slate, which I held at one end, he at the other, with our left hands: he then produced a double slate, hinged and locked. Without removing my left hand I unlocked the slate, and at Mr. D.'s direction, placed three small pieces of chalk—red, green, and grey—inside: I then relocked the slate, placed the key in my pocket, and the slate on the table in such a position that I could easily watch both the slate in my left hand, and the other on the table. After some few minutes, during which, to the best of my belief, I was attentively regarding both slates, Mr. D. whisked the first away, and showed me on the reverse a message written to myself. Almost immediately afterwards he asked me to unlock the second slate, and on doing so I found to my intense astonishment, another message written on both the insides of the slate—the lines in alternate colours, and the chalks apparently much worn by usage.

My brother tells me that there was an interval of some two or three minutes during which my attention was called away, but I can only believe it on his word.[30]

Austin Podmore.

---

[30] Mr. Frank Podmore had been previously informed by me as to the details of the particular methods which I intended to employ in the séance described above. —S.J.D.

The Possibilities of Mal-Observation, &c.

Statement of Mrs. Johnson.

My sisters and I being most interested in the subject of slate-writing and anxious to see something of it, Mr. Davey kindly arranged a meeting at his house. We sat at an ordinary table in a well-lighted room, and writing was quickly produced on the inner surface of one of two slates held firmly together, once by Mr. Davey and myself, at other times by my sisters and Mr. Davey; at first just under the edge of the table, then above, and afterwards on one of my sister's shoulders. This was the more wonderful as we had purchased the slates on our way from the station. Of course between the slates were placed three points of different coloured chalks, after which Mr. Davey asked us in which colour the writing should appear, and it did so in the colour we elected, the slate being covered with writing.[31] We are all quite certain that the slates were never out of the hands of one or other of us, and we are totally unable to account for the slate-writing.

M. Johnson.

[*September*, 1886.]

Statement of Mr. Scobell.

November 25th, 1886.

Dear Sir,—

I had the pleasure of attending a séance given by you some few months ago, and beg to relate what took place to the best of my recollection. First, you produced a framed slate which folded, and upon which there was a patent lock. You opened the slate, cleaned it perfectly free from writing, put two or three pieces of crayon or pencil therein, locked it up, and placed the key in the hands of one of my daughters, who was present. The slate was laid on the table, and the hands of all of us were placed on and around it. You then told us to think of some subject upon which we should like a few lines, and to say the colour in which we should like them to appear. This was left to one of my daughters. You then appeared to be invoking the aid of some unknown person, which appeared to be attended with considerable mental agitation to yourself, and as light scratching was heard, and upon the slate being finally unlocked and opened, two or

---

[31] These words were added by Mrs. Johnson later.—S. J. D.

three lines of writing appeared therein, and they were upon the subject my daughter had lent her mind, and in the colour writing desired by us.

The next thing you did was to solicit us to take out any volume from your bookcase, turn to a page, and fix our special attention on a passage. This I did without your seeing the page or passage. The book was handed to you, and you in a short time told us the right page and right paragraph.

I can only say that my daughters and myself were perfectly astonished with your performance, and had we been predisposed to believe in Spiritualism, we should have been convinced in such belief through your séance, as the whole performance seemed to us a phenomenon incapable of any explanation and not to be produced by any ordinary natural means. —Yours faithfully,

R. W. SCOBELL.

Mr. S. Davey, Jun.

Statement of MR. S. ELLIS.

November 29th, 1886.

Both Mrs. Ellis and myself are pleased to have an opportunity of testifying to the intense gratification you have afforded ourselves and friends on several occasions, both at our house as well as under your own roof, by the psychical phenomena you have exhibited, and we are now as much as ever at a loss to arrive at the *natural means* employed, as at the startling results produced—so astonishing that it is almost impossible to believe even the testimony of two senses. The productions of the "locked slate" fully bear out the foregoing statement.

S. ELLIS.

Statement of MRS. BARRETT

Your wonderful performance on the slate completely puzzled me. I have not got over it yet. Thinking over it as much as ever I can, I am as far off having any idea about it as at first. You say you did the writing, so I suppose you did; but how? That is what I want to know. You gave me a clean slate without a mark or scratch of any kind upon it. I examined it carefully, I sponged it with water, and at your desire I locked it up and kept my eye upon it. When it was unlocked and the slate examined, I discovered, to my astonishment, that it was written all over from top to bottom. I never lost sight of the locked slate, and I never lost sight of you; and as far as I could judge, it was impossible for you or any one present to have done it; yet the wonderful fact remains; the slate was perfectly clean when it was locked up, and written all over when unlocked. This is a mystery, and as I am unable to look through a wooden cover, I cannot imagine a clue to it. Perhaps some of these days you will enlighten me.

Statement of Miss Stidolph.

I have much pleasure in recording my recollections of a séance with Mr. S. J. Davey. His powers are certainly marvellous, and while I have not the very smallest belief in "Spiritualism" or "mediums" of any kind, believing the things so called to be gross deceptions, I was amazed at my friend's scientific skill. Apparently he has no appliances. I was seated with him at a small table when he gave me the following astounding evidence of his powers. He gave into my hands a slate which, when locked, looks like an ordinary box. This box I opened, washed the slate, locked it, and took the key; for some minutes we sat, he with one hand on mine, his other hand on the table. Presently a faint scratching was heard, and continued some little time; when it ceased Mr. Davey unlocked the slate, and lo! it was covered with clear, distinct writing—a letter addressed to myself, and stating if I would wait a little while the writer would go to the Cape and bring me news of my brother. Then I again washed the slate; again it was locked, and again I kept the key. Mr. Davey then asked me to take any volume I liked from the library, to look at a page and remember the number of it. This I did, and again we sat as before. In a few moments the slate was unlocked, when on it was written, not only the number of the page I had thought of, but some of the words which were on the self—same page, and these not ordinary words, but abstruse words, as the book I selected was a learned one. This I considered a most marvellous feat, and utterly incomprehensible. That the scientific researches of my friend will lead to most important results I have no doubt. His aim is to expose deception, and if this object be attained he will benefit society and throw light on a subject which has hitherto been considered to belong exclusively to the "powers of darkness."

E. Stidolph.

I would mention that the shelves from which I took the book contained hundreds of volumes, and Mr. Davey had no idea which I had selected as he closed his eyes and went to the extreme end of the room.

E. S.

November 25, 1886.

Proceeding now to more detailed accounts I shall first quote reports by Mr. J. H. Rait and Mr. Hartnall J. Limmer, of a sitting which I shall call

SITTING 1.[32]

These accounts were written independently, from notes taken during the sitting. Mr. Limmer had had a successful séance with me some months previously, of which he wrote no account.

1. *Report of* Mr. Rait.

On Wednesday evening, the 8th September, 1886, at 7.30, I betook myself, in answer to a previous invitation, to the residence of Mr. S. J. Davey. I had brought with me at his request three new common school slates privately marked by me and of medium size, a box of assorted crayons, and a book to take notes in. Arrived there I was introduced to Mr. Limmer, who with Mr. Davey and myself formed the trio in whose presence the manifestations which I am about to record took place.

At 8.30 p.m. we seated ourselves as shown in the diagram. Mr. Limmer sat directly opposite me, while Mr. Davey sat on my left, the gas burner being directly overhead so as to distribute light equally on all surroundings. Before I begin, however, I will call attention to the following facts.

1. During the whole séance, with but one slight exception, the gas was burning brightly.

---

[32] I have numbered the sittings in the order in which I have quoted them, and have also lettered the chief events described in each, so that where more than one account is given of the same sitting, the reader may easily compare the different descriptions given. The small index numbers in parentheses refer to notes which will be found in the *Appendix*, pp. 192-204.

The Possibilities of Mal-Observation, &c.

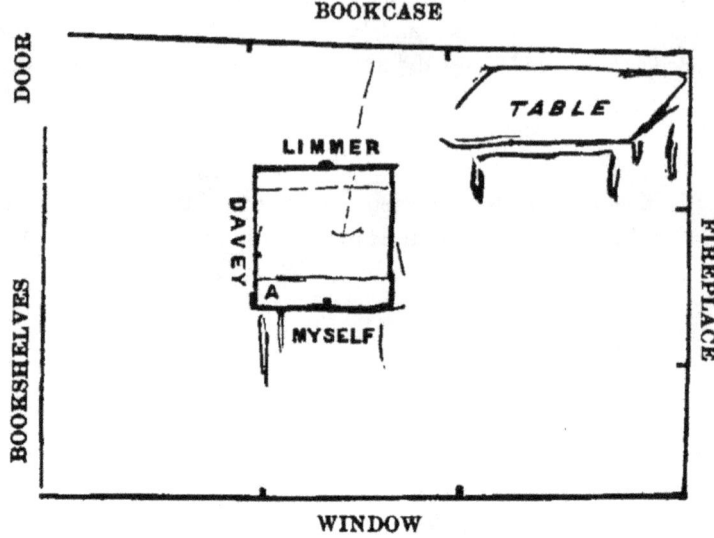

A. This is where the Slate was held.

2. The slates used were the 3 already mentioned and a double one of Mr. Davey's of superior make, with ebony backs and fitted with a lock, which, after having cleaned it and inserted a small fragment of slate pencil, I locked, and at his request put it in the pocket of my coat, where it remained till used. With these slates there could not possibly be any tampering,[1] as during the whole séance they never for one moment left the room.

3. While the writing was taking place under the table, Mr. Davey's left hand was held by Mr. Limmer while his right with the exception of the tops of his 4 fingers was full in my view.

4. The chalks used were my own, wrapped separately in paper, and before the séance had never been taken out of the box.[2]

5. A fact that appears to me most wonderful is that the point of the slate pencil or crayon was always worn and invariably[3] formed part of the last stroke.

At Mr. Davey's request I took one of my new slates, cleaned, wiped it, and placed a minute fragment of slate pencil on its surface, and held[4] it under the table at the corner of the table with my left hand, pressing it firmly all the time. Mr. Limmer held my right on one

Experimental Investigation

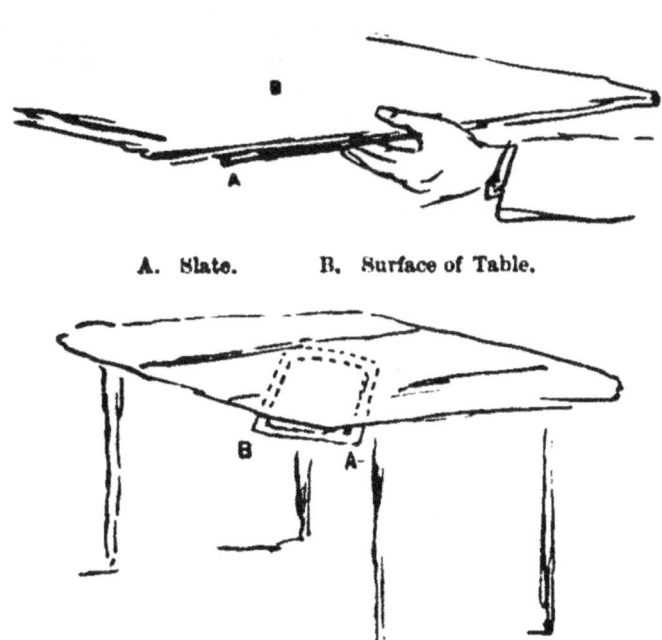

A. Slate.   B. Surface of Table.

The dotted lines show that part of the Slate under the Table.
A. Extremity held by me. B. Held by Davey.

side and Mr. Davey's left on the other, while Mr. Davey also supported the slate under the table with his right; thus it will be seen that a chain was formed by the hands. After remaining a few seconds thus.

[a] Mr. Davey: "Are you there?" No answer.

Mr. D: "Are you going to give us any answer this evening?" A distinct ticking sound was here heard and after 3 seconds or so it stopped and I withdrew the slate; on it was an imperfect scrawl which no one could decipher.

[b] Mr. D.: "We will try again; please hold the slate firmly; engage in conversation; try and not concentrate the thoughts too much on one subject."

Mr. D.: "Are you going to give us any answers this evening or not? Now *do* try." This time the noise of the ticking of a pencil was most distinctly heard as if firmly and deliberately writing. I called Mr. Limmer's attention to the fact and he informed me the writing was distinctly audible to him. I withdrew the slate and on it distinctly written was *Yes*.

[*c*] Mr. D.: "Will some one now ask a question?"

After some thought it occurred to me to ask what o'clock it was at present, there being no timepiece in the room.

Mr. D.: "Will you kindly tell us what time it is?" The ticking was *immediately*[5] resumed. I watched Mr. Davey while seemingly talking to Mr. Limmer, but could detect nothing suspicious in his movements; three distinct ticks were heard and I put the slate on the table and examined it. It was written in the same indistinct hand, and began with a scrawl, but in the middle of the sentence I could decipher "nine" plainly. On asking Mr. Limmer to look at his watch he replied that it wanted a quarter of an hour to 9.

[*d, e, f*] On putting the question "Will there be a war with Russia?" we got the vague reply "*Perhaps*." In reply to other questions the answers obtained were "*try chalk*" (this refers to the difficulty experienced in distinctly writing on a new slate) and "*answer later*."

So far nothing striking had occurred beyond very scrawly writing, and replies which might mean anything; but something better was in store for us.

[*g*] I now suggested[6] a slight variation in the experiment, which both Mr. Limmer and Mr. Davey agreed to. I will mention however that in the right-handed breast pocket of my coat I had placed a sealed envelope containing some questions of a most impossible nature, and which I had written on the afternoon of the 7th September, intending to produce them at the séance with a view to getting them answered; they being all the time in the envelope and their contents unknown to anyone but myself. I determined therefore to put the question, "What does the right-handed breast pocket of my coat contain?"

Requested by Mr. Davey to clean and again privately mark my slates, I did so; and at his request Mr. Limmer and I chose 3 fragments of chalk,—pink, green, and blue. These 3 fragments were placed on the surface of one of the slates. I then placed another slate on the top of this so that the chalks were between.[7] This time the slates were above the table; we joined hands and began talking, the question concerning my coat pocket having been put. It is important to note that during this experiment both of Mr. Davey's hands were in view, also that the writing began almost instantaneously on joining hands. Mr. Davey became very agitated, his hands slightly trembled under mine, and he occasionally gasped for breath as though in pain. (These fits occurred at intervals throughout the séance and always when the writing was taking place, but on no occasion did he move either his hands or feet.) The writing distinctly continued, cool, deliberate, and steady. I could even hear the occasional dashes as in stroking the t's, &c.; it invariably seemed to come, away from Mr. Davey, immediately underneath my fingers. I could almost feel the chalk as it moved along in its weird progress, guided by what mysterious agency I know not.

All at once Mr. Davey said, "Quick!! in what colour will you have it written?" Pink was chosen. This is what appeared on lifting one slate off:—

DEAR SIR,—This experiment is a very difficult one, and we can but rarely repeat it. (In green) You may rest assured that we shall do all in our power to answer (in blue) you this evening, but we are very anxious that you——not——this question (in pink) simply on account of the—— question we will try and answer your question later on—— and the—— endeavour to convince——any test you may suggest. "ERNEST"

The latter part written in pink. Part of the message we could not decipher, and I accordingly cannot repeat it in full. This message occupied about 2 minutes or less in writing, and was on the whole fairly well written.

[*h*] The next experiment was with Mr. Davey's closed slate. After it had been produced from my pocket we laid it on the table locked and with the small piece of pencil inside, joined hands as before and the question was put, "Will the Emperor of Germany live through the present year?"[8] Immediately the writing began, exactly the same as on previous occasions, and when after the space of 4 minutes (about) I carefully unlocked the slate we found the following wonderful message: "My Dear Sirs, —It is a popular error that if we can produce this writing under these conditions we might at the same time have a knowledge upon all questions of a mundane nature. One is apt to forget that prophet seer and prophetess are children all of 'mother guess,' and this rule applies to us. Yet for ourselves we can foresee much to

The Possibilities of Mal-Observation, &c.

happen in in the year *1889*, and to do this we need but carry out the instructions of Bonnet (?) who said, 'Ne vous lassez jamais d'examiner les causes des grands changements, puisque rien ne servira jamais tant à vôtre instruction. 'Your test is a severe one, for we have not the gift of clairvoyance tonight. On VII——we think (or thank) your friend from time to time in explanation of this mystery try your test again later on and we shall succeed. We hope to——" (here the writing ends). This is clearly a direct reply to all our questions, and "the severe test" referred to, points evidently to my *coat pocket's contents*. What the mysterious VII. means I do not know, except that it may have some allusion to the 7th September, the day on which I wrote the questions. This belief is strengthened[9] by the answer we got in trying to find out the writing after the Roman letters VII, later on in the evening, and which read (as much as we could make out of it) *Septem*. This long message was to my mind the most marvellous result of all, and its effect was strongly marked on Mr. Davey, who seemed in a state of great prostration, and called for a glass of water.

[*i, k*] Mr. Davey then placed a slate on two small boxes which rested on the table, thus; 3 pieces of chalk,—blue, pink, and red—were then chosen and placed on its surface (the slate) and over the chalk was placed a tumbler; the gas was slightly lowered, and we were told to say what figure we would like to have drawn. I chose an octagon, Mr. Limmer chose a square. I saw a piece of chalk slightly move and on

lifting the glass we saw two very indistinct marks. We however resolved to try again. This time the red piece of chalk distinctly moved, but very quick. Lifting the tumbler we found this figure which evidently was intended for part of Mr. Limmer's square.

/

[*l, m, n*] I desired[10] after this to have the writing on the double slate of Mr. Davey's continued at the point where it had been broken off, and obtained this result on one of my slates which I held underneath the table and which began immediately. "We hope to see you again —Joey." I was also anxious to know what the VII signified as I have already said before;—on the first attempt we got the answer—"good-bye Joey"—but we were more successful on again putting the question, the result being a distinct "Septe——"; whether, as I have already said, it was intended for September I cannot tell.

As it was getting late (10.30) the séance concluded. In finishing this statement I will add that for my part I am "an outsider," have never before given slate-writing or Spiritualism a thought until Mr. Davey lent me "Psychography" and a copy of *Light* dated 8th November, 1884, and invited me to relate my experiences as they appeared to my senses of sight and hearing only; which I have endeavoured to do in as complete a manner as possible. What the agency is that moves the fragment of pencil I know not; I leave that for the savants. It is a wonderful thing that part of an answer was written in French, a language totally unknown[11] to Mr. Davey. Also that 3 colours were employed in writing another answer. Trickery to my mind is utterly impossible in any respect. How it is all done I cannot tell; my advice to the "sceptics" is "go and judge for yourselves."

JOHN H. RAIT.

10/9/86.

2. *Report of* MR. LIMMER.

On Friday, the 8th September, 1886, I had the privilege of being present at a "Spiritualistic" séance given by Mr. S. J. Davey at his residence Mr. Herbert Rait was the only other person present besides Mr.

The Possibilities of Mal-Observation, &c.

Davey and myself.

The only table used was a small one which Mr. Davey informed us was technically known as a "Pembroke." This table I thoroughly examined, and nothing that could aid Mr. Davey in any way could I discover. The proceedings then commenced by placing a common slate, bought that evening and marked by Mr. Rait, under the corner of the table and supported in that position by the right and left hands of Mr. Davey and Mr. Rait respectively, while I completed the circle by holding their disengaged hands.

[*b, c*] The question "What is the time?" was then asked by Mr. Rait, and after a short interval I distinctly heard writing, but on looking at the slate the answer was not readable: the question was therefore repeated, and shortly after the word "nine" was obtained.

[*d, e, f*] The next question asked by Mr. Rait was, "Will there be a war with Russia or not?" in reply to which we received the word "Perhaps." The same gentleman then asked "Will the Emperor of Germany live through the year?" Instead of receiving a direct reply the words "Try chalk" were found written upon the slate, and on adopting that suggestion we obtained the single word "later."

I may mention here that all the chalk and slates (with the exception of the "locked slate" mentioned later on in this report) used during the evening were brought by Mr. Rait, and had never been in the possession of Mr. Davey.

[*g*] The next test was that of two common slates being placed upon the table, one above the other, the frames of which fitted so accurately that it appeared utterly impossible to insert anything by which the pencil could be put in motion. These slates were previously examined by Mr. Rait and myself. Green, pink, blue and red chalk having been inserted by Mr. Rait, the circle was again formed in the manner before described, Mr. Davey having this time, though, both hands placed upon the top slate. The question, "What does my right hand breast coat pocket contain?" was put by Mr. Rait, and it was agreed that the colour in which the answer should be written should be pink. I distinctly heard the chalk passing rapidly between the slates, and in about two minutes we had the following message before us.

(In pink)

Experimental Investigation

"Dear Sir,

"This experiment is a very difficult one, and we can but rarely repeat it. (In green.) You may rest assured that we shall do all in our power to answer (in blue) you this evening, but we are very anxious that you should not put this question (in pink again) (word not plainly written here) simply on a / c of the (word not readable) question. We will try and answer your question later on, and the (word not readable) endeavour to convince (word not readable) any test you may suggest.

"Ernest."

At this stage of the proceedings Mr. Davey appeared to be rather exhausted, and drank a glass of water.

[*h*] Mr. Davey then produced a "locked slate," which I examined most minutely, and as far as I was able to judge, the surfaces were genuine slate and had not undergone any process of preparation which would aid him in obtaining writing. A small crumb of pencil was inserted, and the slate closed and locked by Mr. Rait. The key was then given into my possession. We then placed our hands in an exactly similar position as before, and Mr. Rait having repeated the question "Will the Emperor of Germany live through the year?" I very soon heard the pencil travelling over the surface of the slate. After the lapse of about four minutes the slate was carefully unlocked by Mr. Rait, and the[12] pencil very much worn was found at the place where the writing ended.

The lines on the first side of the slate ran in a diagonal direction from left to right, but on the second side it was done in the usual manner, i.e., from side to side. The writing was of a very neat character and the majority of the letters were well formed. The following is a copy of the letter.

"My Dear Sirs, —It is a popular error that if we can produce this writing under these conditions we might at the same time have a knowledge upon all questions of a mundane nature. One is apt to forget that 'Prophet, seer, and prophetess are children all of Mother Guess' and this rule applies to us, yet for ourselves we can forsee much to happen in in (the word 'in' occurred twice here) the year 1889 and to do this we need but carry out " the instruction of Bonnet (this name was indistinct) who said 'Ne vous lassez jamais d'examiner les causes des grands changements puisque rien 'servira jamais tant à votre instruction.'

Your test is a severe one for we have not the gift of clairvoyance tonight on VII oz we think (or thank) your friend from time to time in explanation of this mystery.[33]

Try your test again later on and we shall succeed.

We hope

Saw Pencil lie here, on carefully opening the Slate.

[*l, m, n*] The writing having stopped so abruptly, two[13] ordinary slates were placed upon the table in the manner before described, and it was asked by Mr. Rait that the letter should be concluded. Within a period of 15 seconds from the time of asking such question and after completing the circle with our hands, the words "to see you again, Joey," were written.

The two slates were again placed in the same position as before, and Mr. Rait having put an unimportant question, after the completion of the circle as before, I saw upon the slate "Good-bye, Joey"; but on a second trial a scrawl was obtained which looked very much like "Sept. Joey" but it was impossible to say definitely what it was intended for.

[*i, j, k*] The final test to which Mr. Davey was subjected was that of writing under an inverted tumbler under the following conditions. An ordinary tumbler was inverted and placed upon one of the slates brought by Mr. Rait. This slate was raised slightly from the table and supported by two small boxes placed under the ends of the slate. Blue, pink, and red chalk were then placed under the glass by Mr. Rait, and after joining hands, Mr. Rait asked that an octagon should be formed

---

33 This probably refers to some questions which Mr. Rait had written and enclosed in a sealed envelope and placed in bis breast coat pocket and known only to himself. It will be remembered he previously asked "What does my right-hand breast coat pocket contain?"— H. J. L.

with the red chalk. After waiting for a few minutes the red chalk was seen to make two short lines almost at right angles to one another, thus, 7 The same test, after the slate had been cleaned, was repeated, and with precisely the same result. I then asked that a square should be formed by the red chalk, and two sides of it were made almost instantly, and in the colour required. Although looking to within a few inches of the tumbler and seeing the pencil move, I failed to discover anything which could have caused it to do so.

I can only say that the whole thing was totally inexplicable to me, and to the best of my belief it was impossible for Mr. Davey to have produced any of the above results by the aid of trickery, as he did not appear in any way to try to divert my attention either from himself or the slates, and I watched him as closely as it was possible throughout the whole proceedings.

HARTNALL J. LIMMER.

SITTING II.

The following three reports are by a member of the Council of the American Society for Psychical Research, and his wife and daughter. I shall speak of them as Mr. and Mrs. and Miss Y. The reports ought to be specially instructive in consequence of the differences of attitude illustrated by the sitters. Mrs. Y. was unaware, until after her report had been written, that the phenomena were nothing but conjuring. Miss Y. was unaware of this fact during the sitting, but I understand that she was unintentionally informed of the true nature of my performances before she wrote her report. Mr. Y. was aware, before any arrangement had been made for the sitting, of the work upon which I was engaged, and knew that any phenomenon which might occur would be due to my own conjuring powers.

1. *Report of* MRS. Y.

On the evening of September 10th, 1886, I went with my husband and daughter to a room in Furnival's Inn, to witness the slate-writing performances of Mr. Davey. On our way we stopped at a stationer's, and my husband purchased three perfectly new ordinary school slates. We found Mr. Davey to be a young man of manifest intelligence and great earnestness of scientific purpose. He impressed me as being thoroughly honest and above all trickery. He also impressed me as being in a very critical state of health, and I should say the nervous strain of his slate-writing performances was most injurious to him.

[a] We seated ourselves at an ordinary Pembroke table, brought out of the kitchen attached to the chambers belonging to the friend who had loaned his room for the occasion. A piece of chalk was placed on one of our slates, and the slate was held tightly up against the underside of the table leaf by one of Mr. Davey's hands and one of my daughter's. Their thumbs were on top of the table, and their hands spread underneath on the underside of the slate. I held Mr. Davey's other hand, and we all joined hands around the table. I watched the two hands holding the slate without a moment's intermission, and I am *confident* that neither Mr. Davey's hand nor my daughter's moved in the least during the whole time. Two or three questions were asked without any sign of response.[1] Then Mr. Davey asked rather emphatically, looking hard at the corner of the table under which they were holding the slate, "*Will* you do anything for us?" After this question had been repeated three or four times, a scratching noise was heard, and on drawing out the slate a distinct "Yes" was found written on it, the chalk being found stationary at the point where the writing ceased. As my eyes were fixed uninterruptedly on both my daughter's hand and on Mr. Davey's also, and as I certainly had fast hold of his other hand all the time, I feel *confident* he did not write this word in any ordinary way.[2]

[b] This same result was obtained two or three times. But Mr. Davey did not seem to think it was enough of a test, and he proposed that we should try it with the slate on the table in full sight of us all, with a candle[3] burning brightly in the middle of the table.

[c] He gave me a locked slate of his own, which I thoroughly washed and locked myself, and put the key in my own pocket. We then joined hands, and Mr. D. and my daughter placed one hand each on the slate as it was lying on top of the table. Different questions were asked, and we waited some time, but no response came. Mr. Davey seemed to me very much exhausted, and I urged him to desist from

any further efforts. But he seemed loth to do this, and said he would rest a little while, and would then, perhaps, be able to go on. After a short time of conversation, the slates all the while being in full view and carefully watched by me, we again tried it, under the same conditions as before, only that this time Mr. D. requested us each to take a book at random from the shelves in the room, and mentally think of two numbers representing a page and a line, and he would see if he could reproduce it. This also failed of any result, and Mr. D. said he feared he was too tired to produce anything, as he had been very much exhausted by a long and very successful séance the night before. We again begged him to desist, but after a short rest, during which he walked into the next room for fresh air, I thought, he insisted on another trial. The slates still remained all the time in full view on the table. Mr. D. asked my daughter to choose another book, which she did at random, he having his back to her and standing at some distance while she did it.[4] This book was at once tied up and sealed by one of the party, Mr. D. never touching it from first to last. I then held it in my lap, while we joined hands as before, and Mr. D. and my daughter each put one hand on the slate. Still nothing came. Then we changed positions, and I placed my hand on the slate instead of my daughter, giving her the book to hold. During this change she kept her hand on the slate until I had placed mine beside it, and the book was awaiting her on the opposite side of the table, my husband all the while holding Mr. D.'s other hand. I am *confident* that Mr. D. could not possibly have manipulated the slate during this change, for it was in full sight all the while, and our hands were on it, and the book was tied and sealed on the opposite side of the table. A few minutes after this readjustment Mr. D. seemed to have a sort of electric shock pass through him, the perspiration started out in great drops on his forehead, and the hand that was touching mine quivered as with a nervous spasm. At once we heard the pencil in the slate moving, and in a few moments Mr. Davey asked me to unlock the slate. My daughter took the key out of her pocket and handed it across the table to me, and I unlocked the slate, and found it covered on both the inner sides with writing. When read, this writing proved to be a sort of essay or exhortation on the subject of psychical research, with quotations from the book chosen intermingled throughout. I forgot to say that Mr. D. had asked us *all* to choose in our minds two numbers under ten to represent a page and a line of the book, but had finally concentrated his thought on what my husband was thinking. In the writing there were quotations from every page we had any of us thought of, but not always the line; but in the case of my husband the line was correct, but not the page. He had thought of page 8, line 8. The line was quoted from page 3, and Mr. D. said this confusion between 8 and 3 quite frequently occurred, because of the similarity of the numbers. This test seemed to me *perfect*. The slate was under my own eye on top of the table the whole time, and either my daughter's hand or my own was placed firmly upon it without the intermission of even a second. Moreover, we closed and opened it ourselves.

[*d*] After a short rest, Mr. Davey asked us to wash two of our own slates and put them together, with pieces of chalk of different colours between, and all of us to reach across the table and hold them all together. This we did, and then Mr. D. asked my husband to choose mentally three colours he wished used in writing. After all holding the slates closely pressed together for a few minutes, we placed them on the table, and Mr. D. and I placed our hands on them while the rest joined hands. In a few moments the same sort of electric shock seemed to pass through Mr. D., and his hand and arm which were on the slates quivered nervously, and immediately a scratching noise was heard. He then asked me to lift one slate off the other, which I did, and found one side covered with writing in three colours, the very three my husband had mentally chosen. I am perfectly confident that my hand was not removed[5] from the slates for one single instant, and that I never lost sight of them for a moment.

By this time Mr. D. seemed to us to be so much exhausted that we begged him to give up any further tests, but he insisted on trying one more, which was as it proved the most remarkable of all.

[*e*] He placed one of our slates on three little china salt-cellars that lifted it up about an inch from the table. Upon the middle of this he placed several pieces of different coloured chalks, and covered them with a tumbler. Then he told my husband to form a mental picture of some figure he wished to have drawn on the slate under the glass, and to name aloud the colour he would have it drawn in. He thought of a cross, and chose aloud the blue colour.[9] I suggested that blue was too

dark to be easily seen, and asked him to take white, which he agreed to. We sat holding hands and watching the pieces of chalk under the tumbler. No one was touching the slate this time, not even Mr. D. In a few minutes, Mr. D. was again violently agitated as with an electric shock, which went through him from head to foot, and immediately afterwards we saw, with our own eyes, each one of us, the pieces of chalk under the glass begin to move slowly, and apparently to walk of their own accord across the space of the slate under the tumbler. My husband had said just before that if the piece of red chalk under that tumbler moved, he would give his head to anyone who wanted it, so sure was he that it could not possibly move. The first piece of chalk that began to walk about was that very red piece! Then the blue and white moved simultaneously, as though uncertain which was the one desired. It was utterly astounding to all of us to see these pieces of chalk thus walking about under the glass with no visible agency to move them! All the while Mr. D., whose hands were held on one side by myself and on the other side by my husband, seemed to be on a great nervous strain, with hot hands and great beads of perspiration. When the chalks stopped moving, we lifted the tumbler, and there was a cross, partly blue and partly white, and a long red line marking the path taken by the red chalk! We were impressed by this test beyond the power of words to declare. The test conditions were perfect, and the whole thing took place under our eyes on top of the table with no hands of anybody near the slate. This was the close of the evening's performances.

 Upon reading over my account I see that I have put the leaving of the room by Mr. D. in the wrong place. It should have been just before the writing on our slates with coloured chalks instead of just before the writing on the locked slate. But in either case the slates were all the time in full view on the table with the rest of us who remained behind. I consider the test conditions to have been perfect throughout, and see no possible explanation for the very remarkable phenomena that occurred.

Mrs._____

September 14th, 1886.

2. *Report of* Miss Y.

The exhibition was given in Mr. Hodgson's sitting-room, a medium-sized room with a large square table in the centre, covered with a cloth. Mr. Hodgson and my father and mother and I were in the room, seated around this table, when Mr. Davey entered. He looked at the table, and said it would not do. So we pushed it aside, and a Pembroke table was brought in in its place; I do not know whether it was the property of Mr. Hodgson or whether Mr. Davey had brought it with him. It was quite bare, and we placed on it a candle and three single slates which my father had bought at a shop on our way to Furnival's Inn. We also had on it a bowl half full of water, containing a sponge to wash the slates with, and a cloth with which to dry them. On the large table was a lamp and on the mantelpiece were three candles, so the room was quite clearly lighted. We sat in a circle around the table, my mother next to Mr. Davey, then my father, then Mr. Hodgson, then I by Mr. Davey.

[*a*] Mr. Hodgson brought us a little pasteboard box, in which were a number of small pieces of chalk of different colours. I chose two of these and placed them on one of our slates. We had all previously written either our names or our initials on that side of the slate. Mr. Davey slipped the slate under the edge of the table, I holding on to it all the time, and we held it flat under the table with our thumbs above the table. I held the slate very firmly against the table, and I am sure I did not relax my hold once. After waiting some time and asking various questions, we heard, or seemed to hear, the chalk moving on the slate. We drew the slate out,[10] and on it was written "Yes," which was an answer to our last question.

[*b*] We again put the slate under the table, and, in order to be sure that nothing had been written on it, I half slipped it out again and saw that it was perfectly clean. After some more waiting, my father asked when we were to sail for America. The chalk again squeaked, and on drawing the slate out we found "the 18th" written very indistinctly. This happened not to be the date, which was the 15th. I forgot to mention that while we were waiting for the writing we had all joined hands, and talked on indifferent subjects. Just before the writing came Mr. Davey grew very quiet and writhed with his arms. His left hand, the one not holding the slate, pressed very hard upon the table. Mean-

while, my father and mother had changed places, because it was indicated in some way, by the writing on the slate, I think, that we were not all seated in our proper places.

[c] After these experiments, Mr. Davey seemed very much exhausted. But he drank some water, and insisted on going on to try the experiment with the books. He had previously brought out his own little slate, a double one made of ebony, with a silver lock on it, He said he would use this in the experiment. We saw that it was clean, and then one of us locked it. I think my mother put the key in her pocket. Then we each chose a book at random from a large book case at the end of the room, in which the books were arranged apparently without any system. I can't remember whether we left the slate on the table while we chose our books, or whether one of us held it.[11] When we sat down at the table again, I put my hand on the slate and leaned my elbow on my book. Then Mr. Davey asked us each to think of two numbers under 10, one for a page in our book and one for a line on that page. This we did, and sat waiting for sometime with our hands joined, but no writing appeared on the slate. Mr. Davey seemed so exhausted that we determined to give up the experiment, and I put my book back in its place. We sat as before around the table, discussing the failure of the experiment. Finally Mr. Davey started up and said, "We must try it with one book alone. Will you choose one, Miss____?" I supposed that he asked me to do it because my seat was nearest to the bookcase. I got up and went to the bookcase. Mr. Davey stood by the table with his back to me. That latter fact I feel as if I remember most distinctly. I mention it to show that I chose my book at random and was not influenced in my choice by him. As I came back to the table he said, "Do not let me see the name of the book." But as I did not understand what the trick was to be, I forgot his injunction and placed the book on the table at my right hand. Mr. Davey was on my left. However, I only left it there for a minute or two, and I am sure that I either looked at it or held it the whole time, so that he could not have opened it without my knowledge. Mr. Hodgson brought me some string and this I passed around the book several times and tied with four knots. Then my mother sealed the knots and took the book. She did not hold it in her hand the whole time, but either let it lie in her lap or sat on it.

But Mr. Davey could not have taken the book or opened it without her knowledge, as he sat perfectly still and we saw everything he did. We opened the double slate, and after we had seen that it was perfectly clean, I put some chalk in it, locked it, and put the key in my pocket; and I also kept hold of the slate during the experiment. Mr. Davey asked us each to think of two numbers as before. Finally he asked us to write them down on a slate. I wrote mine on one of our own slates so that he could not possibly see what I had written, and I placed it on the table away from Mr. Davey, and leaned my elbow on it.

I think the others did the same with the other slates. To my remembrance,[12] some of us watched the locked slate all the time while we were writing.

After a few minutes Mr. Davey asked my mother to change places with me. This we did, but I did not relax my hold of the slate until she had her hand on it. She gave me the book, and I sat on it, and we again joined hands. Mr. Davey's right hand was on the slate which was on the table, and my father held his other hand. Mr. Davey said that the experiment was too difficult while we all thought of different numbers. So he asked my father to think of his numbers, and the rest of us not to think of ours. After a little while the chalk inside the slate squeaked a good deal. I took the key out of my pocket, and one of us, my mother, I think, unlocked the slate. Both sides were covered with writing, all of which I will not quote. Then we cut the string of the book, having found the seal untouched. We opened the book at my father's page and line, 8 and 8, but there was nothing there that was quoted on the slate. So we looked at page 3, line 8. There we found this line, "The greenest grasses Nature laid." On the slate was this sentence, "The *greenest grasses* will be laid by *Nature*." It was not in quotation marks, although three of the words were underlined. On my mother's page, 8, but not on her line, 4, was the title of a poem, "The House of Clouds." This was on the slate, underlined. On Mr. Hodgson's page, 7, but not on his line, 9, was the title, "The Deserted Garden." This was also on the slate, underlined. On my page, 1, we found nothing that was quoted on the slate.

[d] After this experiment, we put aside Mr. Davey's slate and took two of our own. We cleaned them, and placed on one a number of little pieces of coloured chalk. The second slate was put on the first

Experimental Investigation

one, and my mother and Mr. Davey held it above the table. Mr. Davey asked my father to think of three colours. We joined hands once more, and in a little while we heard writing between the slates. When we took one off, on the under one was written:—

In *red*, "We are very glad to be able to give you this."
In *white*, "We can do more yet."
In *green*, "Good-bye."

My father had thought of red, white, and *blue*. We could not be sure by the night light whether the "good-bye" was written in green or blue. But there was a piece of chalk on the slate that looked much more blue than the piece with which the "good-bye" was written.

[*e*] After this we tried one more experiment. Mr. Davey placed one of our slates on two little Japanese salt-cellars, made of china covered with wicker, and one common glass salt-cellar. On the slate we put a number of pieces of coloured chalk, and over this a clean tumbler. Meanwhile Mr. Davey took the candle from our table, and put it on the other one. This made the light a little less distinct, but it was still very good. Then Mr. Davey asked my father to draw a figure and write a colour on his double slate. My father made a star, and wrote "Red." This he showed to all of us except Mr. Davey, after which he locked it, and put it in his pocket, and gave me the key. We joined hands, and Mr. Davey's hands did not touch the slate or the glass at all. We sat for sometime, without any results.

Finally, we gave it up, and my mother and I put on our cloaks to go home. But before we left the room, we decided to give it one more trial. We thought that perhaps we had made too complicated a figure, so we unlocked the slate, and rubbed out what my father had written before. Then I drew a cross, and he wrote "Blue." We sat down at the table and watched the chalk with the closest attention. Mr. Davey's arms shook violently, and once when the chalk began to move, he snatched his hands off the table. But my father and mother did not let go of them. While we were waiting for the chalk to move Mr. Davey seemed very much excited, and he asked my father what colour he had written. My father said "Blue," whereupon my mother said, "What a pity you did not say white. It would have been so much easier to see the white move." So my father said, "Very well, let it be white." At Mr. Davey's request, both he and I kept our minds fixed on white, and on the figure I had made on the slate.

Finally, the red chalk actually did move across the slate. Mr. Davey snatched the glass off. I expected to see a red cross, but the red had only made a slight short mark. There was a long white mark, and across it, near the top, was a green mark. But the green mark was very small, and not at right angles with the white mark, nor did it extend as far on one side of the white mark as on the other.[13]

That is all that happened, as nearly as I can remember. It took place two evenings ago, the 10th of September. The only curious thing I noticed about Mr. Davey was the odd way in which his arms, and sometimes his whole body, writhed, especially just while the slate-writing was going on. At the time, I was convinced that they were not tricks, although I had no other explanation for them.

September 12th, 1886.

3. *Report of* Mr. Y.

On September 10th, 1886, last evening, with my wife and a daughter of nineteen years of age, I availed myself of an invitation to see the phenomena of slate-writing at the rooms of Mr. Hodgson, No. 1, Furnival's Inn, London.

The "medium" was introduced as Mr. Clifford [Davey, see p. 85], a gentleman of known social standing who had never accepted any pecuniary returns for his performances.

Mr. Hodgson's room was, throughout the conference, lighted by four candles and one shaded lamp, there being no moment of obscuration of light through the whole evening. The large heavy table in the room was, at Mr. Davey's suggestion, substituted by a smaller one with two folding leaves, the table ordinarily in use in Mr. Hodgson's breakfast-room. The three slates were wood bound, about 8 x 10 in. size. They were purchased by me at a cost of threepence each, from a stationer in Holborn, on my way to the conference. Their surfaces were very rough, requiring much washing with a sponge and dry rubbing before they were smooth enough for use.

Mr. Davey had two much larger slates with false black card surfaces, showing how persons were often imposed on by professional mediums; but of these no use whatever was made.[34] He also had a small silver-mounted ebony framed locked slate. I shall distinguish this by "D.'s slate," and the ones I bought by "my slates." In a card paper box were a number of very small pieces of pencil of six different colours. The whole apparatus has now been described.

[a] Mr. Hodgson sat opposite to me, on my left sat my wife opposite to my daughter, between whom and myself Mr. Davey was placed. A pencil was put on one of my slates, which was sustained under the edge of the table by my daughter's fingers at one end and by Mr. Davey's at the other, their thumbs being all the time in sight on the top of the table. The hands of all five of the party were then joined, and soon we heard a sound like that of a pencil. My slate was slid out from under the table by my daughter. At the first and second examination nothing was on the slate, and it was washed afresh, and soon the word "yes" was found scrawled on the upper side of the slate as an answer to some indifferent question.

[b] This was once repeated.

My daughter was sure that her hand had pressed the slate to the under side of the table during the whole time. It is to be noted that during about a minute of the time of waiting Mr. Davey seemed seized with violent nervous contractions of his face, arms, and hands, which appeared to draw largely on his strength. They were succeeded by a feverish condition of the surface of his hand.

[c] He next requested us to select, each one, a book from Mr. Hodgson's library. We did so, and he proposed to give us the contents of lines, selected by page and line by us, in D.'s locked slate. But nothing came. He then proposed to try it again on a single book, and my daughter, leaving him at the table, replaced[14] on the shelves the book she had first taken down, and took at random a copy of Mrs. Browning's *Poems*, Second Series. 1880. London edition.

He requested us to mentally choose each one a page and line by numbers below 10.

    I selected page 8, line 8,
    My wife page 8, line 4,

---

[34] The remark which I believe I made was that the slate with a false flap was often put forward by professional conjurers as an explanation of the slate-writing tricks performed by "mediums."— S. J. D.

Mr. Hodgson[15] page 7, line 9,
My daughter page 1 line 9,
and we each one told aloud[16] the page and line selected.

During this time and previous to the announcement of the selection, a pencil had been put into D.'s slate, which had lain on the top of the open table throughout. It was then locked, and the key placed in my daughter's pocket. My wife's hand was then placed upon D.'s slate as it lay on top of the table, and it certainly so remained until, after hearing the sound of writing, my daughter produced the key, and we opened the slate.[17] My choice had been page 8, line 8, and while the others conversed Mr. D. bade me fix my mind intently on these figures during a space extending probably to 10 minutes. Then my wife unlocked D.'s slate, and found the following words written in legible hand. The italics are as in the original.

"How far you remain, oh unbelieving ones, from the goal of your endeavours. It is not through the domain of Physical Phenomena, but through the empire of the soul's dominion, that man must enter upon the higher paths that stretch away into the Divine. 'The Kingdom of God is within you.' Seek not to entangle the brightness of your soul in the Labyrinthine *Maze* of the physical world *which will prove* to be merely a *House of Clouds*, and will leave you more forlorn than a *Deserted Garden*, where not even the *greenest grasses* will be laid by *Nature* to sanctify her right. We men a " (here it ends with a long irregular line, such as might be made by a pencil rolling).

On referring to the book which had been lying in our sight tied up and sealed, we found on page 8 the heading to be *The House of Clouds*, being the words italicised in the slate-writing. On page 3 (which, Mr. D. said, the power, whatever it was, might have mistaken for 8 as being like it) and line 8, we found "the greenest grasses Nature," words also italicised on the slate.

The other words, "*which will prove*," in italics were not in the pages of poems indicated. There seems to have been no attempt to give any words from the lines selected by the other three persons.

[d] We next placed small pencils, in six colours, between two of my newly bought slates, marked by ourselves with our names written in pencil, without removing them from the top of the table, and the hands of some of the party were laid upon them for some minutes,

after which they were held up in the hands of two persons. I had been asked to choose the colours in which the writing should be made. I mentally chose red, white, and blue, but did not tell my choice. After holding Mr. D.'s hand for some minutes, with my mind strongly fixed on these colours, the slates were opened, and we found, in the order I had mentally selected:

(Red) "We are glad to be able to give you this."

(White) "We can do more yet."

(Blue) "Goodbye."

[*e*] The last performance was the moving of the pencils on my slate under a glass. Two Chinese cups were taken from the mantel. On these the slates rested, with pieces of pencil cut by ourselves from the ends of coloured crayons of various colours, on its top, under the glass. Upon D.'s lock slate, which he could not see, I wrote "Blue" as the colour which should be used, and my daughter made a cross for the figure to be written. We then locked D.'s slate, which I placed in my pocket, while my daughter put the key into her pocket. We joined hands as before, but there was no movement of the pencils. He took the glass off, threw off the pencils, and put them back again in a group, near the centre. Again we joined hands. I selected white for the colour, at my daughter's outspoken suggestion, and Mr. D. became spasmodically excited, trembling greatly. Soon the red pencil began to move in full view of all of us. and soon began again to move. The distance traversed was probably an inch. I did not see the other pencil ends move, though they might have done so.

It stopped

Upon removing the glass we found traces in white, green, blue, and red, but scarcely the figure traced by my daughter on D.'s locked slate.[18]

Having in my boyhood practised legerdemain I am able to see how a large portion of the publicly performed tricks are, or could be, done; and surprising as were the very able performances of Mr. D., I could tell how most of them could be done[19] without anything beyond a probable mind-transference of thought intensely concentrated on my part. But I stood in the same relation to the performances which I could not unravel, in which others stood to those which I could perform or explain. I cannot therefore make my own discernment the limit of the "natural," and say that the performances undiscovered by me are "supernatural." I am surprised when I find that those versed in conjuring, when they reach the limit of their own ingenuity and knowledge, ascribe to supernatural causes what is beyond their ken.

Mr. Davey is a gentleman, I believe, incapable of intentional deception. He makes no statement of his performances beyond the fact that they are phenomena to be accurately observed. They exceed in their apparent supernaturalism the displays usually made by paid mediums. I believe that a full explanation of his methods would "fire a shot heard round the world" in almost every civilised community where the phenomena of so-called "Spiritualism" are perplexing, and often madden, true and good people.

Whatever may be the real psychical phenomena which lie behind or alongside of the supposed revelation from spirits, they should be separated from the often proved deceptions of paid mediums. And I can conceive of no more effectual step towards this than for some one who excels these magicians upon their own chosen field, to frankly tell the world how it is all done. Scientists then would have the ground cleared for accurate investigation, and, more important still, millions might be saved from the delusions of deceitful "mediums." For the atrocious wickedness of deceitfully trespassing by fraud upon the most holy of all human relationships, the sacred regard for the dead, it is difficult to find strong enough terms to express our contemptuous reprobation.

Experimental Investigation

SITTING III.

The sitting described by Mr. Legge in the following letter took place on the same evening as the one described in the foregoing three reports, September 10th,1886.

Report of MR. LEGGE.

12, Mitre Court Chambers,
Temple, E.C.
12th September, 1886.

MY DEAR HODGSON,—

I said I would let you have a straightforward account of what I saw in your chambers on Friday evening; here it is, written while my recollections are distinct. I had been sitting with Hughes, who told me you had a séance on. We were talking on various subjects, and never touched that of the séance going on in the next room, so that in fact I soon forgot all about it. Presently however the door opened, and you came in with Mr. _____, whom I had never seen before, and behind him were his wife and daughter. A little embarrassed by sudden introductions, I passed out into the entry to your rooms, and there saw for the first time Clifford (the name, is it not?) [Davey, see p. 85]. In a few minutes the _____s left, and I went into the room where the séance had been. There I was introduced to Davey, who suggested, after some talk on what had gone before, that I might like to see something.

[*d*] I jumped at the proposal, and as a preliminary took Davey's locked slate, opened it, and cleaned it (or rather cleaned it again, for it was clean already), then, having slipped in a bit of pencil, locked it, and put the key in my pocket, keeping the slate also in my hand or under my arm all the time afterwards.[1]

[*a*] Next I chose one of your slates, cleaned it, and, at Davey's request, having put a fragment of pencil upon it, slipped[2] it under the table, so as just to cover it with the slab, the pencil of course being on the upper surface. The table was a plain deal one, and I satisfied myself that there were no projections on the under-surface which could leave a mark. We then formed contact, Davey's right hand being partly on my left and partly on the slate, i.e., where his hand overlapped mine which held the slate pressed against the table. The faint sound of

The Possibilities of Mal-Observation, &c.

writing was distinctly audible, and when it ceased I drew the slate out. There were merely indistinct scrawls upon it.

[*b*] But it appeared that I ought to have asked a question to myself, and so extracted an answer from the slate. So after I had cleaned the slate I thought of a question. I acted precisely as before.[3] The sound of writing was again heard, and the slate, when I drew it out, bore clearly and distinctly written the word "yes"—the final letter of which was done in particularly admirable style. The question[4] had reference to some doubtful increase in my official salary, and I am bound to say I was as much delighted as astounded by the mysterious writing.

[*c*] The next experiment was the placing of 3 bits of coloured chalk on the table, and of a clean slate (selected and placed by myself)[5] over them. I put my hand on the slate, Davey his on mine, and we joined contact. Again we heard the sound of writing, and when I lifted the slate there was written large and neatly in the coloured chalks (three lines or so in each colour) this message:—"Don't you think I've done enough for you tonight I'm tired Joey." I noticed the chalks seemed worn, showing signs of work, just like the little bit of pencil in the previous experiment.

[*d*] After this Davey asked me to write a question in chalk on one of your slates. While I was writing it he asked for a drink of water, and you pointed to a corner of the room where there was some. He went there and when he came back seemed to have forgotten his request for he now asked me for the locked slate, which I had latterly put in my pocket.[6] I brought it, out, placed it on the table, set my hand on it, Davey his on mine, and joined contact just as before with Hughes and yourself, Hughes holding my right and your left, you Davey's left. Then we heard the same writing sound, very faint this time, and after a considerable interval I was told to take up the slate and unlock it. Taking the key out of my pocket I did so, and saw written on both leaves a long message, precisely as I give it: — "If you don't believe in spirit power after this you are not worth the attention of any honest medum (*sic*) Joey." After this, Davey, who seemed pretty tired, had to rush off to catch a train. I should add that the room had been in full lamp light all the time, the lamp being placed on a side table thus throwing a certain light under as well as over the table we were sitting at. Also that the contact was

not continuously perfect, for I remember that Hughes and yourself occasionally left one hand free for a short time, and lastly that there was no enforced silence.

The above are the facts as detailed as I can give them; I offer no comment on them for indeed I can't. Though I had heard of such experiments before, personal experience was entirely new to me, and has left me in immense perplexity.

If you see Davey, please thank him for his kindness in troubling after an already long sitting to give me some specimens of his "craft," or whatever name one can give so nameless a faculty. I am also sincerely obliged to Hughes and yourself. —Yours,

J. G. Legge.

SITTING IV.

The next three reports are by Mr. Padshah, and Mr. and Mrs. Russell. Mr. Padshah had had some previous experience with a well-known medium, but had not been convinced that the phenomena which he had witnessed in the medium's presence were not the result of trickery. His account of his séance with me shows that he was in some respects a careful observer, and that he was still more careful in recording his remembrances. He was not informed until after he had written his report, that the phenomena were due to conjuring. Mr. and Mrs. Russell, however, knew before the sitting that they were about to witness conjuring performances. They came on an unexpected visit to Mr. Hodgson, and on learning that I was about to give a sitting, requested permission to be present. I was perfectly conscious of the fact that they were both using their best endeavours to discover my exact *modus operandi*. And although Mr. Russell failed to detect any of my writing processes, he correctly observed and remembered some of my manipulations with slates above the table, which, it will be seen, entirely escaped the observation or remembrance of Mr. Padshah.

The Possibilities of Mal-Observation, &c.

1. *Report of* MR. Padshah.

1, Furnival's Inn, London.
Sept. 15, 1886.

This evening in Hodgson's room we had a séance with Mr. Davey; Mr. and Mrs. Russell, Mr. F. S. Hughes, Hodgson and myself being the party. Before sitting I had some interesting conversation with Mr. D. about the results usually got by him and some which I had with Eglinton. Mr. D. in course of the conversation told me he was very anxious that his results should be tried and watched like those of any professional medium, and indeed, his subsequent proceedings were very agreeably contrasted with those witnessed at the professional séances. There was every apparent desire to get the conditions named by members of the party, and to see that results were obtained under those conditions. I had suggested in our preliminary conversation how important it might be to get my own name—not surname—which no one except myself in the room knew. The slates on which we desired the writings were three of them Hodgson's, three I had bought this evening at Lilley's, Cambridge, and one Mr. D.'s own double slate. I regret that desiring to add some friends to the party, I had left the rooms to call upon those friends, and during that interval, Hodgson and Mr. Hughes being busy, we necessarily could not keep the slates in our eye for a short interval during which Mr. D. was in the room.

Well, we commenced, I sitting all the time next to Mr. D., except once, when Mrs. Russell and I interchanged places, with no advantage; and so we resumed the original order.

[*f*] There was full light on every corner of the table; two of my (?) slates, one washed by myself, the other[1] by Mr. D., were put very nearly in the centre with a number of small chalk-pieces between them of different colours—(five in all, I find now on inquiry from Hodgson—red, blue, green, yellow, white—but which I was *not* sure of, then, not having noticed them).

[*a*] Under the table with the frame projecting on Mr. D.'s side, was a single slate, also mine, I believe, and washed by I do not know whom, Mr. D. supporting it on his side by the four fingers underneath, and the thumb over the table in sight of all; his left hand joining with that of Mr. Russell's right, Mr. R.'s left with Mr. Hughes' right,[2] Mr. Hughes' with

Mr. Russell's, and Mrs. Russell's left with my right, all resting either on the table, or otherwise always in sight; and my left supporting also the slate just the same as Mr. D. Between the slate and the table were put successively chalks and a small pencil, the chalks being crushed, and therefore given up.

[*f*] Mr. D. and Mr. Russell often put their hands on the pair of single slates.

[*g*] Mr. D.'s double slate, *not* washed,[3] I believe, that I can remember, but locked up by myself carefully, with the key always in my pocket or on my RIGHT hand near Mrs. Russell, never out of my view, was in my charge, generally being behind my back.

[*a*] For some time there seemed to be no result, Mr. D. telling us that he felt no "go" in the thing, and asking me if it was not due to my undue scepticism. Of course I assured him that my failing was rather in the reverse direction. The conversation was generally on Spiritualistic subjects, being mostly a good-humoured discussion of the experiences of some Spiritualists. On Mr. D.'s asking me to select a particular colour of chalk to write between the two slates,[4] I suggested white; but we never got it.

[*b*] He then wished me to fix my mind on a particular number. I selected five (5), and drew an image of it before my mental eye. The number we got was, however, 6; and I must say, that but for the horizontal stroke, I myself would be unable to distinguish often between my 5's and 6's. Mr. D. then asked if there were going to be any manifestations — the answer[5] was legibly "Yes."

[*f*] Then we[6] asked for a writing on one of the pair of slates, of mixed colours, mine being blue, and Mrs. Russell's selection red. *Sometimes* I think we all put our hands on the pair of slates and then both Mr. D.'s hands were in full view, and there could be no mistake of what they were doing,—viz., that they were shaking sometimes with great force, at the same time that his teeth were chattering.

[*d, e*] However, before any writing came there as asked for, we had first a message on the single slate "Wait," and at another time, I noticed (without any clear sound of writing as was unmistakeable during the two previous cases) and I believe nobody had observed it before I drew their notice—a message on one side of the slate, "Try the (?) chalks."

[*f*] Well, now we all concentrated our attention on the pair of slates very nearly in the centre; and I thought, as requested, of two numbers, 5, 7; Mr. D. very shortly after a deal of shaking of his hands, at length said that we might see the slates. There, to my surprise, I beheld a message forsooth, in two coloured pencils,—blue and red, which I copy below.

(Blue Pencil):

"We are very pleased to be able to give you this writing under these conditions, which must or ought at least to the ordinary mind do away with the possibility of it being produced by ordinary means."

(Red Pencil):

"If you will be kind enough to wait patiently you may rest assured we will do our best to do more for you."

"Earnest."

[*c*] I forgot to say that before this writing appeared, on the large slate, instead of the numbers we wanted, we got written[7] "Boorzu." Now this as it happens is the original Persian, the modern corruption of which is my initial name. This would be extraordinary except that it might have happened by accident, and also I had not time enough to see the last "u" before the word was wiped off by Mr. D.

[*g*] Then we tried to get some results with books, but as it appeared to me Mr. D. had read almost every book in Hodgson's library, it was not easy to select one to preclude the hypothesis of thought-transference. So we attempted to get numbers again, and I concentrated my attention on the same two previous numbers (5, 7); we soon got the 7 on the single slate, but instead of the 5, we got "Think Book." Mr. D. desired me to think of one; my mind was unsettled between *The Brain as an Organ of Mind*, by Bastian, and *International Law*, so to avoid any interference with the conditions, I pitched upon the periodical, *Mind*. Mr. and Mrs. Russell having left us, we all concentrated our attention on the double closed slate, which, on opening at frequent intervals, we had found unwritten. The key was *now* in my pocket, *that* is certain, for on seeking to open it, I found it entangled with the coppers in my waistcoat pocket. The double slate was also undoubtedly locked, for I carefully locked it myself. I mentally, as before, concentrated my attention on getting the word "Mind" written within. After some time Hodgson said he heard the sound of writing, and on opening it we

found the slate full. The following is the text: —

"This phenomenon is not Spiritualistic, nor is it the projection into objectivity through the higher faculties unfolded by the abnormal issues of human developments—'Mediumship'? Yes. But mediumship of WHAT? Do you think you could appreciate if we were to tell you? Ah no! The Spooks of one, the Adepts of another, the transcendental Egos of another, and the fourth dimensions of a fourth, are but the frantic struggling dreams of the dark and ignorant present human race who have not acquired the possibility of CONCEIVING even an approximation to the real solution.

"Your own predominant desire is to *explain*, but for these and kindred facts, it will be ages before the loftiest soul can touch the true *theory*, as we find it exhibiting no distinct changes of form, and if impossible with one or more vibration.

"The Brain AN organ of Mind, ha! we laugh."

This completes the text. I opened the slate myself, and I found some scratches made by the pencil over the writing. Also the facet seemed to have worn out a little by writing. After this we made some fruitless efforts at getting something, but we could not, and in a very short time we adjourned. As the table round which we sat was removed, Hodgson pointed out that it was beyond suspicion,—a fact which I had omitted to notice.

[*f*] How came, now, the writing between the pair of slates, and in the closed double slate? About the former, it is certain that the slate on which the writing came was one of the three I had purchased that evening at Cambridge; as was attested by its size corresponding with the two others marked, and also by the shape of the frames, and the cracks in them noticed by Hodgson. I confess I do not remember even after such a brief lapse of time, whether I had examined[9] the two slates *not* washed by me, and found them unwritten. I imagine I must have, for otherwise it would be very stupid; and, besides, if there had been any writing it would not have escaped the notice of Mr. Russell, who seemed to be particularly careful. Besides, we constantly looked to see if there was any writing there. Of course, a conjurer of ordinary pretensions could deceive on the last point. There might be writing on the bottom surface of the lower slate, while we could observe only the three upper surfaces, if so many. Before we saw the writing there,

Mr. D. gave a push,[10] and though I am almost sure that it was I who removed the upper slate, and found the writing there, I am afraid I cannot be certain. Indeed, I doubt if I can with any confidence assert whether the writing was on the lower surface of the upper slate, or the upper surface of the lower slate, even if I was certain that it was not on the lowest face. When I remember that Mr. D. is deliberately anxious to be tried by no other than a conjurer's standard, and also that I have omitted to notice things so elementary, and yet so essential, even some of them actually suggested for my observation by Mr. D., I regret I did not ask some one else of the party to observe and act. For it is evident that if I did not see the slates clean on *all* the surfaces before commencement, my testimony becomes absolutely valueless. But now suppose that we have satisfaction on these heads, still it may be considered possible that the writing may be precipitated by chemical means. Whether, if the writing disappears under the influence of water, the chemical theory may still hold, of course I cannot say. But if so, it is curious that Mr. D. could push the slates at a particular moment; and before that none of us could notice, in that full light, any formation of letters, or gradual precipitation, that I can see. Besides, Mr. D. could barely have had time enough to tamper with the slates. He told me himself that he had observed them lying. He had almost 40 minutes to himself, with little intervals, when Hodgson would come in. He might during that interval have written out all the first message, without using a chemical; in that case we are all guilty of gross negligence which it is ridiculous to credit my colleagues[11] with. But he might have also used a chemical; only he could not have foreseen the opportunity of my going out; and as everyone is supposed to bring his own slates, why he should carry about chemicals with him it is difficult to see. In this connection I may also observe that Mr. D. remarked to me during our conversation after tea, how great the temptation is for the occultist to be fraudulent; when pecuniary remuneration is not the object, "the good of the cause" is supposed to justify them, and it may not be unjust to add,—the desire to make people talk about them is not altogether a factor without influence. Just imagine the temptation in Vanity Fair of an * * the guest of princes and emperors, and having the great honour of a recommendation from the first of living Englishmen—Mr. * * ! But it is only fair to Mr. D. to say that he holds this justification, he says, in great abomination. As for the selection of colours being blue and red, and turning out so, it seems to me quite natural, and it may not be without significance that the white writing with chalk we asked for never came. Besides, there is nothing in the matter itself which may not have been written beforehand, indeed it was not what we had wanted. Now, though I point out my own defects of observation, it is only to show how little really my testimony is worth except for points of confirmation; and I hope I shall be able to remedy them next time.

[g] Somewhat different is the case of the double closed slate. I do not remember it to have been washed; but there never was any writing on it except a scratch occasionally, whenever I opened it, with the exception of the last time. As I opened it myself I think I could easily have observed any gradual precipitations. The reference to "Brain as an organ of Mind" is not altogether without significance. It is also evident that Mr. D. must have minutely studied the time it takes for complete precipitation; or that the whole precipitation takes place simultaneously; or that the phenomenon is undoubtedly genuine. The theory of mere writing without a chemical and then bamboozling me would be really contemptible.[12]

[c] The reading of numbers was not a failure; but it was not convincing. "Boorzu," however, was remarkable.

On the whole, I myself strongly incline more towards the genuineness of the phenomena than the reverse; but I cannot disguise it from myself that that is largely due to a previous impression gathered from Mr. D.'s results with others which were read out to me. If I get the same things next time with *my own* double-slate, and a pair of slates that have never left my sight, I think I should be justified in being convinced of something abnormal.[13]

2. *Report of* MR. RUSSELL.

[16, Somerfield Road, Finsbury Park, N.]

On Wednesday evening, September 15, I was present with my wife at a slate-writing séance given by Mr. Davey. We sat in the private sitting-room of my friend Mr. R. Hodgson, at No. 1, Furnival's Inn. Besides Mr. Davey, Mr. Hodgson, my wife and self, there were present

Mr. Hughes (another great friend of mine) and ... Mr. Padshah. I had never seen either Mr. D. or Mr. P. before. We sat round an ordinary deal table. Mr. P. was on Mr. D.'s right hand, I on his left. On the table were 3 or 4 single slates which Mr. P. had brought with him, and a double slate fitted with lock and key belonging to Mr. Davey.

[g] As soon as we were seated at the table Mr. D. washed the double slate with sponge and water, and then handed it round for inspection. As we expressed ourselves satisfied that it was perfectly clean, he placed a small piece of ordinary crayon inside, locked it and gave it to Mr. Padshah to keep. Mr. P. having put it on his own chair behind his back.

[a] Mr. D. took one of the single slates, washed it clean, put a small piece of crayon on it and placed it under one corner of the table, holding it there with his right hand (thumb in sight on the table, four fingers out of sight below), Mr. P. holding it in the same manner with his left hand. We then joined hands and talked,[14] waiting for the sound of writing. After some minutes Mr. D. brought up the slate, but there was nothing on it.

[f] He then put some small pieces of chalk on one of the other slates lying on the table, covered it with another slate, and said he would try to get some writing there if we would choose the colours we would like it in. Mr. P. chose blue and my wife (at my suggestion) red.

[a] Mr. D. then replaced the single slate under the corner of the table, holding it as before, but again several minutes passed without any result. He then asked my wife to change places with Mr. P., which she did, holding the slate with her left hand as he had done. But again, after several minutes, there was no writing.

[b] Then my wife and Mr. P. took their old places, Mr. D. once more put the slate under the corner as before, and asked Mr. P. to think of some number under 10, saying that he would try to get it written for him. He then said aloud: "Please say whether we shall get anything tonight," soon after which Mr. P. declared he heard the sound of writing; whereupon the slate was brought up, and the word "yes" and the number "6" were found upon it.[15] Mr. P. said he had thought of 5, but explained that he made his fives in such a curious way that they might easily be mistaken for sixes.

[c] Mr. D. now said that a start having been made, more success

might be looked for, so the experiment was repeated, the slate being brought up at intervals of from 5 to 10 minutes. The first time it had the letters BOORZ[16] upon it, which Mr. P. explained were the first five letters of his Christian name which was in Persian written BOORZU. Neither Mr. Hodgson, Mr. Hughes, my wife, nor myself had ever heard of this name before, but I did not quite understand whether Mr. D. had or had not heard it from Mr. P. before the sitting began.

[*d, e*] Next time there was the single word "Wait," and a little later the words "Try Chalks." We accordingly concentrated our attention on the two slates with the chalks between them, which had been left lying on the table.

[*f*] Mr. Davey and Mr. P. each placed a hand on them, and we completed the circle. From time to time Mr. D. opened the slates, but for a long time there was no result.

[*g*] Presently he got up and went to the bookcase, saying he would try to read something from a book. He asked Mr. P. to go and choose one. Mr. P. did so (taking the locked slate with him), and suggested several books, to all of which Mr. D. objected on various grounds. Finally, however, a volume of Swinburne's poems was selected and placed on the table, Mr. D. saying he would try to get a reference in the locked slate to any particular page and line below 10 Mr. P. might choose. But though the slate was opened two or three times, no writing was found on it.

[*f*] In the meantime, Mr. D. had once more examined the two slates where the coloured chalks were, but finding nothing, had placed them side by side, and carelessly, as if in a fit of absent-mindedness, had taken the chalks from the slate which had been at the bottom, and placed them on the other. He had then put them together as before, except that the original position of the slates was reversed, the old bottom one being now at the top, and the old top one at the bottom. Presently, asking Mr. P. if in a former sitting with Eglinton the medium had not got some writing on his shoulder, he took up the two slates and placed them on Mr. P.'s shoulder, but in less than a minute took them off, reversing[17] them as he did so, and replaced them on the table. The old bottoms late was now once more at the bottom, and the old top one at the top, but each slate had been reversed, so that the two sides which had originally been turned to the table were now turned up. In

a few minutes, Mr. D. had a sort of convulsion, Mr. Hodgson and Mr. Hughes said they heard sounds like writing, the slates were opened, and there, on the lower one, was a message, half in green,[18] half in red (nearly the colours chosen by Mr. P. and my wife), expressing a hope that we should be satisfied with writing given thus, under such excellent test conditions. Mr. P. remarked that he had asked for blue, and that the colour given was green; and then, on being asked, said he could not see how Mr. D. could have produced this writing by ordinary physical means, and then my wife and I left.

I am writing this account without notes,[19] on the morning of Friday, September 17th.   J. RUSSELL.

My wife and I have written our accounts independently, but I have since read through hers, and find I have omitted to say that there was a good light in the room.

3. *Report of* MRS. RUSSELL.

[16, Somerfield Road, Finsbury Park, N.]

I was present with my husband at a séance given by Mr. Davey to Mr. Padshah at Mr. Hodgson's rooms in Furnival's Inn, on Wednesday night, the 15th inst.

There were six of us present. We sat round a small deal table, which had a drawer at each[20] end. The one my end was empty. I did not examine the other. Two lamps were in the room, and four candles, one of which was on the table. Mr. Padshah sat next to Mr. Davey and I next to Mr. Padshah.

[g] He began by cleaning the inside of a locked slate given him by Mr. Davey, who having chosen and put inside a small piece of chalk, desired Mr. Padshah to lock the slate and keep it in his possession. Mr. Padshah locked it and put it behind him in the chair he was sitting in, and the key in his pocket.

[a] Mr. Davey then took a small ordinary slate, and a small piece of slate pencil with no points, asking Mr. Padshah to first clean the slate himself on both sides. This being done they both held the slate under the edge of the table with the fingers on the slate and the thumbs on the edge of the table. We then all joined hands, and sat

talking for some time. Once or twice Mr. Davey took out the slate to examine, but found no writing. He then asked me to change places with Mr. Padshah, and hold the slate, which I did. Once or twice he took out the slate whilst I was holding it, and once there was a zigzag pencil mark on it which was not there before, but no writing.

[*b*] Mr. Padshah then took the slate again. We still went on waiting, and taking out the slate to look at. Twice, some white chalk that Mr. Padshah had chosen was crushed when we looked at it. Mr. Davey then bent his head close to the table and asked in a loud voice, "Tell us if we shall have any manifestations tonight or no; only one word Yes or No." After waiting again Mr. Padshah said he heard the sound of writing. On looking, "Yes" was found written on the slate. The letters were very uneven and scrawling. Mr. Davey then asked Mr. Padshah to think of a number, and a figure 6 was given instead of a 5 which he had thought of. But Mr. Padshah explained it by saying that he usually made those figures very much alike, and it would be easy to confuse them.

[*d, e*] After waiting again the single word "Wait" was found, and a little time afterwards "Try chalks" in the same bad writing (so bad that we turned it first one way and then another to make it out) with a very imperfect figure 8 that Mr. Padshah had been thinking of. Mr. Padshah himself discovered this last just as Mr. Davey was putting back the slate under the table.

[*g*] Nothing was yet found in the locked slate.

[*f*] Mr. Davey then put in several pieces of coloured chalks between two slates which had been lying on the table all the time, with one piece of pencil inside, and he and my husband placed their hands on it.

[*c*] On again taking up the slate under the table, a curious word appeared written on it which we could not read, written in much better characters, but which appeared to me to be a foreign word. On Mr. Padshah's looking at it, he exclaimed "Why it is my own name Boorzu, which I am hardly ever called by!" No one at the table knew it was Mr. Padshah's name, Mr. Davey being positive that he had never heard it before, and indeed neither of us had. Mr. Padshah then reminded Mr. Davey that he had asked him to ask his name before tea, which Mr. Davey said he had forgotten.

The Possibilities of Mal-Observation, &c.

[g] Nothing having been written between the two slates, Mr. Davey then asked Mr. Padshah to go to the bookcase and choose a book. He brought one and put it on the table, but Mr. Davey objected that it was too big. I think it was a book of Spencer's. Mr. Davey then went to the bookcase with Mr. Padshah and helped him to choose a book, saying it must be a small one, and in large print, that a good clear, large print was of the most importance. Mr. Padshah, on going to the bookcase, took the locked slate with him. They brought back a volume of Swinburne's poems, Mr. Davey opening it here and there, and observing that the worst of it was he knew that particular book very well. Mr. Padshah then thought of a page under 10, but no writing was given. I then went to the bookcase for a book, and brought back *Aurora Leigh*, which, on Mr. Davey's seeing, he said it was the same as they had had two or three nights before, and it would not do.

[f] He then decided to give up the book test altogether, and concentrated all his attention on the two slates on the table. He asked Mr. Padshah and myself to choose a colour that we would have the writing in. Mr. Padshah chose blue, and I[21] red. There were 3 or 4 different small pieces of coloured chalks[22] in the slates. Mr. Padshah and myself then held our hands over the slates with Mr. Davey and my husband. Mr. Davey became very intense, saying we *must* get some manifestations that night. Mr. Padshah said that perhaps they would not write on the table between the slates, that although they had said "try chalks," they did not say on which slate. We waited some time without any result. Once Mr. Davey put the slates on Mr. Padshah's shoulder, asking if Mr. Eglinton had not tried him in that way. He replaced them after a few seconds on the table, and turned them over to look inside, but nothing was found. At last Mr. Davey became more intense, and after a kind of convulsive shaking, he turned open the slates once more, and, with some excitement, showed us one whole side covered with even good writing, half in green and half in red. I cannot remember what it was exactly, not having taken a copy. But the green was something about giving us a good manifestation that night, and the red about waiting patiently. This last being in my colour, struck me as a curious coincidence, as I had been the most impatient all the evening. Then Mr. Padshah again unlocked the locked slate, but found nothing, and after our waiting some time longer, Mr. Davey suggested

we might perhaps be too many, as he had seldom had such bad results in so long a sitting. As my husband and I wished to get home, we then left, it being past 10 o'clock, and we began soon after eight. Mr. Davey proposed going on with the sitting after we had left, with what results I do not know.

I am writing this account from memory, without notes, on Friday evening, September 17th.

Bessie Russell.

SITTING V.

Previous to my sitting with Mr. Block he had been informed that my "phenomena" were not due to the agency of "spirits," and he was exceedingly sceptical as to the occurrence of any phenomena at all under such conditions as had been described to him.

Report of Mr. A. S. Block.

October 30, 1886.

Dear Mr. Davey,

Few of the persons who have witnessed your extraordinary performances can have done so with more impartial minds than I and my young son, Alfred, did. He, a youth of 16, perfectly ignorant of the whole subject of Spiritualism, mediums, or psychical science, with eyes quick to discern every movement of hand or body; I, calmly observing what I *saw* without desiring to theorise or account for the same, or the way in which it was accomplished.

Having heard of what you were doing I was curious to witness myself your performances, and you kindly gratified me by giving me what I suppose you would call a séance. To my own disappointment, and I fear to your own inconvenience and perhaps greater strain of mind in consequence, I had but half-an-hour with you, having to catch my last train home.

You, my son, and I having adjourned to the library, sat down at a small ordinary table with folding flaps, when you produced several slates and a small folding slate with hinges and patent lock. Giving me the latter you asked me to thoroughly sponge and wipe it, and placing a very small piece of pencil between the two slates, I locked them and gave the key to my son, and placed the slate in my right hand pocket, being the side away from you.

[a] You then handed me an ordinary slate which you requested me also to well sponge and wipe and put a mark in the corner of each side, which I did. Then, putting a small piece of pencil in the middle of the slate you placed it—or slid it—*under* the corner of the extended flap of the table, placing the fingers of your right hand under it, and your thumb on the upper side of the table, and your left hand on the table; I placing the fingers of my left hand next and touching yours under the slate, and thumb on the table, and with my right hand holding the left hand of my son. In a few seconds you said, "Will you ask a question?" when I asked, "What shall I be doing this time tomorrow night?" In about 3 or 4 minutes a slight scratching was to be heard, and you slid the slate from under the table, and only a mark of an illegible word was to be seen.

[b] The slate was again sponged and wiped by me, and again replaced by you in the same position as before—when you, either as part of the performance or in fun, evinced some impatience and demanded an answer to my question, and in a few minutes scratching was again heard, and on withdrawing the slate from under the table, the word "Reading" very legibly written, was on the slate.

[c] You then took two slates which you handed to me to sponge and wipe as before, which I did, and placing 3 or 4 small pieces of coloured chalk, which you placed between the 2 slates, which were placed on the top of the table, you asked my son to take a book from the bookcase, to think of a page without letting you know either the book or the page thought of, and keep the book in his possession. Then asking him in what coloured chalk the writing should appear—he desired it should be in red—you placed both your hands firmly on the upper slate; I placed both mine, and my son did the same, all of us pressing on the slates firmly.

Waiting a few minutes, you again manifested impatience and excitement at the little delay, when we soon after distinctly heard a scratching between the slates, which when looked at, the upper slate was found covered with writing, in red chalk as desired. The writing was apparently an extract of some kind, but unfortunately the opportunity of testing its accuracy was lost as my son omitted to think of a page.

[*d*] Although the time at our disposal was but a few minutes—a quarter of an hour at most—you kindly performed another trick, which was writing between the locked slates. As I have said, these were handed to me by you at the commencement of our sitting, were sponged and wiped by me, a piece of pencil placed between the two slates—locked by me, and key handed by me to my son and the slates placed in my pocket, so that it was manifest you never had any touch or handling of these locked-up slates. Asking me to unlock them I did so and found them in the same condition as when I placed them in my pocket. I, however, again wiped them with the sponge—you replaced the small piece of pencil, I locked them together again, handing the key to my son, and handing you the slates thus locked. These you placed on the top corner of the table, placing both your hands upon them—I and my son doing the same. In about 3 minutes, at most, you began to press energetically upon the slates, when we heard very distinctly a slight scratching between them. You called my attention to the sound, lifting your hands, called my observation to the fact that when you did so the sound stopped,—being again audible when you replaced your hands. In a few seconds taking away your hands, you asked me to unlock the slates, which I did and there saw writing in a good flowing hand—not in your style I observed, on the whole of the upper, and on part of the lower slate. I read the first few lines, which were that it was hoped I had enjoyed the entertainment.

I much regret my hurried departure.

In the above memorandum, I have repeated I believe faithfully what I saw.

Yours faithfully,
Alfred S. Block.

After receiving Mr. Block's report I asked him the following questions:—

1. Kindly say on which side the writing appeared when the slate was held against the table, viz., was it on the lower side where my fingers were or upon the upper side nearest the table; also when you grasped the slate

with me against the table do you remember if you held it firmly or not?

2. Did I endeavour to distract your attention from the slates?

3. To the best of your belief were the slates devoid of writing when you examined and marked them, and did either Alfred or yourself observe the slightest opportunity for my writing on them by ordinary natural means?

Mr. Block replied on November 6th, 1886, as follows:—

1. The writing was on the side of the slate nearest the table, and as you held the slate and I also held it very tightly against the under side of the table flap, it appeared to me to be impossible for you to have touched the pencil or that side of the slate on which the writing appeared.

2. You certainly did not appear to endeavour to distract my attention from the slate—quite the contrary.

3. To the best of my belief and as far as the evidence of my own and Alfred's eyes could be relied upon the slates were all perfectly devoid of writing or marking before the performance, in addition to which as I have said I well sponged and wiped the slates myself and marked them before you received them from me.

I may also state that neither Alfred or I observed the slightest opportunity for your writing on them by ordinary natural means.

A. S. Block.

SITTING VI.

Mr. Ten Brüggenkate had discussed with me some of the literature relating to "slate-writing" phenomena, including some controversy concerning my own performances as "A., the Amateur Conjurer," but I had carefully refrained from making any statement myself concerning the exact nature of my phenomena until after the sitting.

Report of Mr. B. J. Ten Brüggenkate.

November 30th, 1886.

It was my good fortune to witness last night some of the most interesting feats of what appeared to be conjuring that I have ever seen. I had previously had several conversations with Mr. Davey upon the subject of Spiritualism and slate-writing, and last evening when alone with him at his house he volunteered to give me a séance.

The room was a well lighted library, the table at which we sat was an ordinary somewhat old-fashioned Pembroke table, and the slates used were of the common school type, as well as one small folding slate fitted with hinges and a Chatwood lock and key.

[a] Mr. Davey gave me the locked slate and asked me to examine it carefully, which I did and failed to find any trick or anything of the kind about it. The "medium" then asked me to write a question upon the slate, to place a small piece of pencil between the two, to lock it up and put both slate and key in my pocket. I did this in Mr. Davey's absence, he having been called away, for a moment. Mr. Davey then took one of the ordinary slates, and placing a splinter of pencil upon it we both held it close under the table, and after a lapse of a few minutes got some writing upon it, the writing I remarked at the time being in an opposite direction to Mr. Davey. Mr. Davey then returned to the locked slate, *which had been in my pocket all the time,* and upon placing this slate upon the table, very faint scratching was heard, and a complete and full answer to my question was returned. What was to me most extraordinary was, that Mr. Davey did not know what question I had asked, and yet the answer was definite and complete.

[b] The next performance was even more wonderful. I took two common slates, thoroughly cleaned them, and placed some pieces of red chalk between them, and we kept our hands firmly upon them; in a short time faint scratching was heard and upon lifting the top slate I found it to be covered with writing written in a diagonal direction across the slate, the writing again appearing in an opposite direction to Mr. Davey, *i.e.*, as we sat opposite one another it appeared as if I had written it.

[c] The last experiment was only partially successful. Mr. Davey asked me to choose a book from the shelves, unknown to him, and to sit upon it in order that it should be invisible to him—then to write a number upon a slate; I wrote "*five*"—then to think of a number; I thought of "*seven.*" The locked slate was again put upon the table, scratching was heard, and upon opening it I found averse from page 8 line 4 of the book I had chosen, written distinctly upon the slate. I wish it to be observed that I did not fix my mind attentively upon the num-

ber "seven" I had thought of—my attention being called off by some remarks of Mr. Davey; also that Mr. Davey did not know the book I had chosen, so that I quite fail to see how he could produce any writing from the book. This ended the séance, and I am at a loss to conceive how the writing can possibly come upon the slate. There was not a chance of Mr. Davey being able to get at the slates during the performance. When I placed the two open slates one upon another with the red chalk between them, I made the remark that if writing was produced upon either of them I should be ready to believe anything— for they were covered with my hand directly they were on the top of each other and were never moved until writing appeared.

B. J. TEN BRÜGGENKATE.

SITTING VII.

The next two reports are independent accounts of a sitting held on December 1st, 1886. Mr. Venner was introduced to me by a friend in 1885, in order that he might witness one of my performances. He had previously given the question of Spiritualism some thought, and had been present at several séances given by a professed medium. Since that time he has had frequent sittings with me, in company with his own friends. Mr. Manville and Mr. Pinnock I met for the first time at the sitting here recorded.

1. *Report of* MR. ROBERT VENNER.

Séance, December 1st.

On Wednesday, December 1st, my two friends, Mr. M., Mr. P., and myself attended a truly remarkable slate-writing séance given by Mr. D. at his own house. Neither Mr. M. nor Mr. P. have had any previous experience in slate-writing séances. I have been present at something like a dozen of Mr. D.'s; the first of the series must have taken place nearly a year and-a-half ago. At no séance, at which I have been present, have I heard any theory advanced by Mr. D. to account for the production of the phenomena, and he has always strictly guarded himself from any claims to the assistance of the supernatural. I consider that this disclaimer places him in a disadvantageous position, as compared to that of mediums claiming similar results as the work of spiritual agency. In the first place, it debars him from imposing numerous most convenient conditions on the investigator; in the second, it deprives him of much prestige, which cannot but assist the performer; in the third, it prevents him from pressing into his service bad spirits, atmospheric conditions, &c., &c., to account for mistake or failure.

The room in which the séance took place is a small one, and is used as a library; it was well lighted by a couple of gas burners. The table at which we sat was of such a size that all four of us could conveniently join hands when seated; it had two flaps. Before the commencement of the séance we made a thorough investigation of its under-side. The slates employed were all, with one exception, ordinary school ones; no German parchment was used. The exception was a handsome book-slate, cased in black wood resembling ebony, and furnished with a lock. The two halves of the slate fitted very exactly together when closed and locked. The approximate outside dimensions were five inches by eight.

All the slates belonged to Mr. D., whom I shall in future designate as the medium; we brought none of our own. The medium also provided a box of crayons, mostly either red or green, a sponge, a duster, and a glass of water.

Mr. M. objected that the contents of the glass might contain chemicals; we therefore had it emptied and refilled.

The order of sitting was as follows:—Mr. P. and I occupied positions on the medium's right and left hands respectively, Mr. M. sat opposite to him.

[d] At the request of the medium, Mr. P. wrote a question in the book-slate (I shall call this slate A in future); he then locked it and pocketed the key. Neither Mr. M. nor I knew the nature of the question at the time. The slate was left for some minutes upon the seat of an arm-chair, but was subsequently transferred first to Mr. P.'s coat, and then to the table at which we sat. Mr. M. suggested aside to me that we should fix a hair in such a manner to the outside of the slate that it could not fail to be broken if the slate were opened. I thought the suggestion a very good one, but we were not able to put it into execution, no gum being forthcoming, nor any opportunity presenting itself

of distracting the medium's attention.

The medium showed and explained to us a means commonly employed in producing slate-writing by fraud.

Experiment No. 1. [a]

Ordinary slate taken, marked by Mr. P. and myself, and then held beneath the table-flap by Mr. P. and the medium. We got no result during the next half-hour, and Mr. P. and Mr. M. therefore changed places. After a considerable interval the sound of writing audible, and the word "yes" found written; the writing was weak and straggly. As nothing further occurred for some time, the original order of sitting was resumed.

Experiment No. 2. [b]

The medium requested Mr. M. to next ask a question. Mr. M.'s question was something to this effect:—"I had the pleasure of an introduction to a lady last night, I do not know her address, and I should be much obliged by its production." After a considerable pause the word "Marylebone" written.

Experiment No. 3. [c]

Two ordinary slates taken, cleaned by us, but not marked, pieces of red and green chalk introduced between them, the slates then deposited in front of the medium in full view, and about four or five inches from the edge of the table and from the medium's body; the medium rested one of his hands on the upper surface of the top slate, and my hand reposed on his.

After a pause the sound of writing distinctly audible; this continued for about 15 seconds, then the medium remarked, "What a pity I forgot to ask you what colour you would have it in." Mr. M. suggested green; sound of writing continued for about five seconds longer, then ceased. On the removal of the top slate, the bottom slate was found to be completely covered with writing. The writing ran in diagonal lines across the slate; the writing was upside down with respect to the medium; the writing was firm and distinct in character. The first three-quarters of the message were written in red, the last quarter in green; its substance was as follows:—

"We perceive that you possess powers of a very high order, but you have not done what is right for their development. Success can only be obtained by industry, patience, and study, and is not this true as applied to all branches of human affairs? Why should a man be entitled to the assistance of astral angeloids simply because he sits at table and thinks of nothing at all? Ah, no; should you indulge in further investigation with a professional psychic."

This is the end of the red message, the remainder is in German, and written in green. I am not a German scholar, and I shall not give the message in the present report. I understand from Mr. P. that in construction and idiom it is perfectly correct.[35] During the occurrence of the writing, as also before it took place, I watched the medium narrowly, but I could obtain no clue to the means employed. As the writing had been accompanied by some very convulsive spasms of the medium, Mr. M. inquired if these were beyond his power to control. A perfectly frank answer in the negative was returned.

Experiment No. 4. [d]

The medium and Mr. P. placed their hands upon slate A, which had remained insight in front of the latter since the commencement of the séance. The sound of writing audible almost immediately. Mr. P. opened slate, and we found the question he had written, together with the accompanying answer.

Question. "Give me my name in full if you can?"

Answer. "We are sorry we cannot do this for you, Mr. Pinnock; perhaps we may be able to do so later on."

---

[35] Mr. Pinnock wrote a report, but requested me not to publish it unless it was a correct account of what occurred. I may, however, quote the following passage, which I believe to be accurate, with reference to the above incident. "At this point Mr. Davey had asked us if we should like to have the rest written in a different chalk (we had put a red and a green piece on the slate); we assented. I at the same time thought to ask Mr. Davey to let the remainder be written in German, but I did not express this wish aloud. To our great astonishment the first part was written in red chalk, and the next in green, the green writing being in German." This might be described as a communication given to the sitter, in answer to his mental request for a language unknown to the "medium."—S.J.D.

The writing was firm, and distinct in character from that of some of the other messages.

Experiment No. 5. [e]

The medium requested each of us to take a small handful of chalks out of the box on the table. Mr. P. took 11, Mr. M. six, and I three. The medium divided the three chalks I had selected between the other two. We had previously agreed that Mr. P.'s number should represent a page, and Mr. M.'s number a line, of some book to be chosen mentally by one of the party, the medium promising to endeavour to reproduce on the slate the line so determined. In the present case it was of course the eighth line of the 12th page.

The medium requested me to choose a book. I accordingly left the table and walked up to a small case containing, at a rough guess, 60 volumes. I had already selected one of these, when Mr. M. raised the objection that, as I was a personal friend of the medium, it would be a better test if the selection fell either to him or Mr. P. The medium acquiesced.

The slate A was cleaned, and a fresh fragment of pencil introduced; the slate remained in full view with one of the medium's hands resting on it. Mr. M. rose and noted a volume mentally. The sound of writing audible.

The message, on examination, proved to be an address to Mr. M., but contained no quotation from the book he had chosen. I had not time to make a copy of the message in full, but the commencement was as follows:—

"You, who have studied the question of electricity, can the more readily appreciate the wonder of these performances. We think you"——

The medium seemed angry at the appearance of this message, which had no bearing on the question asked, and expressed a desire that we should try a second time. Mr. P. was therefore requested to select a book.

On Mr. Manville asking the reason of the non-success of the experiment, he received the answer "muddle" written on an ordinary slate.

Experimental Investigation

Experiment No. 6. [*f*]

Mr. P. selected a volume mentally without removing or even touching it;[36] he then returned to his place. Two ordinary slates taken, placed together beneath the flap of the table, and held by Mr. P. and the medium. The slates were not specially marked by us, but Mr. P. informed us that the traces of former messages on them offered an easy means of identification. Writing audible. On examination the following message found—

"The difference in this respect Shakespear."

Mr. P. went up to the bookshelf, opened the volume he had selected, and handed it to Mr. M., who found line 8 of page 12 to consist of the following words:—

Line 8. "The difference in this respect between Shake—"
Line 9. "spear and Beaumont," &c., &c.

We informed the medium that he had only been partially successful. Slates held a second time under the table by Mr. P. and the medium. Words "and Beaumont" written.

Slates held under the table-flap for the third time. The omitted word "between" written, and "Shake" instead of the whole word Shakespear. The message was now perfectly correct. The character of the writing in the above messages was weak and straggling.

Experiment No. 7. [*g*]

As we were in doubt as to some of the words written in the message commencing " We perceive, " &c., Mr. M. requested the medium to try and reproduce them.

Two ordinary slates taken, cleaned, and laid on the table in full view.

Almost immediately the sound of writing, and the words "perceive" and "human" written. These were the words in debate. We also got the meaning of certain German words written, the translation of the sentence being, "The weather will change tomorrow." This likewise proved to be the correct rendering.

---

[36] Mr. Pinnock wrote to me on December 14th, 1886:—"In my report I omitted to state that I selected a book mentally without of course telling anyone which one I had selected."—S.J.D.

Experiment No. 8. [*h*]

The medium tore off half a sheet of letter-paper bearing the address of his house; this he gummed to the surface of an ordinary slate, a fragment of lead pencil was put on the paper, and the slate then transferred beneath the table-flap, and held by Mr. P. and the medium. Writing immediately audible. At our request the slate was exposed before it had ceased. To the best of my remembrance the slate could not have been beneath the table-flap for more than 20 seconds. On examination we found the following message written in a hand which bore a much greater resemblance to the medium's than any of the others. Its purport was as follows:—

"D. has not got the mystic instrument up his sleeve or his left hand trousers pocket; we give you this information for the benefit of the SKEPTICS. We do not profess to be possessed of powers out of the range of ordinary human beings, yet we are anxious nevertheless to show you that we can at times give evidence of an intelligence apart from our friend D., and we shall be pleased to try any tests you may devise.

"Mr. V., we are anxious to communicate with you in reference to your relative, Sir R. * * * although of course."

Here I suppose the examination had caused the message to break off short; along pencil mark running from the last letter of the final word seemed to justify this supposition.

This brought a very interesting séance to a close.

In conclusion, I may remark that in addition to the before-mentioned slate-writing séances with Mr. D. I have also sat at a couple of dark séances for materialisation. I can offer no explanation of the phenomena which took place.

ROBERT F. VENNER.

2. *Report of* MR. E. MANVILLE.

2nd December, 1886.

My friend Mr. Venner asked me to accompany him and another friend of his last evening to see Mr. S. J. Davey, who, he said, would show us some phenomena that would probably astonish us. I willingly acquiesced, being not only anxious to see the phenomena (of the nature of which I had been informed), but also to try if I could in any

Experimental Investigation

way observe the means utilised to produce the effects. I may mention I had not seen Mr. Davey before this evening, neither had Mr. Pinnock (Mr. Venner's other friend), but Mr. Venner had known him for some time.

Mr. Davey received us in a small library, probably containing some 300 [over 1000.—S. J. D.] books, and during the whole evening gave me every assistance to examine everything used.

[*d*] I first of all examined a small double slate about eight inches by five inches; this consisted of two slates, each let into an ebony back; the ebony backs were hinged together on one side, and there was a hasp and lock on the other side. When the slates were folded together and locked, the two slates were face to face, with just enough room between them for a "crumb" of slate-pencil locked in between them to move about freely. The slate was washed quite clean with a sponge and water, and dried with a cloth, and then given to Mr. Pinnock to write a question on one side. This he did, and then locked the slates together, retaining the key.

Mr. Davey now brought forward a table, which I examined carefully. It was an ordinary table on four legs, with a flap on each side; it was made of wood about half-an-inch thick; there was one drawer under the table, which I removed altogether, and which was left out all the evening. After this was done, there was nothing about the table which could conceal anything, and had anything been concealed about the table, as far as I could see it must have been in the thickness of the wood.

Mr. Davey then showed me some ordinary slates, in wooden frames. These I helped him to wash and dry. We then took our seats round the table. I was facing Mr. Davey. Mr. Pinnock was seated on Mr. Davey's right hand, and Mr. Venner on Mr. Davey's left hand.

[*d*] Mr. Davey asked Mr. Pinnock to place the locked slate under his (Mr. Pinnock's) coat and then button up the coat.

[*c*] We now took three slates, on one of them we placed three fragments of crayon, two of which were red, the other green, we then covered up this slate with another and left them on the table in full view.

[*a*] On the third slate we also put a piece of crayon and then held the slate underneath one flap of the table which we put up for the purpose. Mr. Davey's fingers were under the slate and his thumb on the

table; Mr. Pinnock's fingers and thumb were in the same position. Mr. Venner held Mr. Davey's free hand with one of his hands and one of my hands with his other. I held Mr. Pinnock's free hand with my other. I have omitted to say that we all three wrote our initials in different corners of the slate before it was put under the table. We sat in this way talking and smoking for some time, twenty minutes to half an hour I should say, nothing whatever occurring. At last Mr. Davey asked me to change places with Mr. Pinnock. This I did and thus had one of my hands on the slate. Mr. Davey now said, that in the manner usual at séances we would ask questions of an imaginary being; and he said, "Are you going to do anything to night, Joey?" After a short pause he repeated the question, and then I felt the slate vibrate as if being written on, and could hear a scratching noise; we took the slate from under the table-flap and saw the word "yes" written over Mr. Venner's initials, and I particularly noticed that the writing was *towards* Mr. Davey, and upside down to him, and in all we saw afterwards this was the case.

[*b*] I now asked a question as to the whereabout of a person at that time, not knowing the answer myself; we waited for some time without any result, when Mr. Davey asked me to again change places with Mr. Pinnock.

[*d, c, g & c.*] I did so, and Mr. Davey told Mr. Pinnock to place the locked slate on the table beside the two slates we had left face to face, and we also lifted the uppermost of these two slates and found the slates still quite clean, with the three pieces of crayon between them. We again waited some time with no results; meantime, having a discussion as to mediumship of different people, and then Mr. Davey asked if I were a medium. After a pause I heard vigorous scratchings on the two slates left face to face on the table and on which Mr. Davey's arm was resting, his two hands being engaged, one in holding the slate under the table flap, the other in holding Mr. Venner's hand; the scratching lasted roughly under ten seconds, and I expected to see a dozen words or so, and was therefore amazed to discover, when the top slate was lifted, that the underneath slate was covered with writing from corner to corner, and also the writing was not straight across the slate, but was across it diagonally; three-quarters of the writing was in red, the other quarter in green, and *no crayon was left*. We read through the writing, a copy of which will appear in Messrs. Venner and Pinnock's report, and found that the part in green was in the German language and characters; about five words were illegible, and these, later on in the evening, we asked for, and obtained them, and still later in the evening we asked for whom the writing was intended, when my name "Manville" was written.

[*d*] Mr. Davey now put his hand on the locked slates which had been left on the table since Mr. Pinnock took them from under his coat; we heard scratching inside. Mr. Pinnock then took the key from his pocket and unlocked the slate and handed it to me. I for the first time saw the question written in it, with an answer below; the question was "Give me my name in full if you can;" the answer was "We are sorry we cannot do this for you, Mr. Pinnock, perhaps we may be able to later on."

[*e*] Mr. Davey now said he would endeavour to get a given line on a given page of a book written for us. Mr. Venner therefore looked over the titles of the books ranged on the shelves and selected one mentally, without touching it with his hands; at this moment I suggested it would be better if I were to select the book, as I did not know Mr. Davey at all, whilst Mr. Venner did. Mr. Davey acquiesced. I selected a title in order to decide what line and page we should select. I took a pinch of crayons from a box, Mr. Pinnock doing the same. On counting, mine came to 6, Mr. Pinnock's to 11, Mr. Venner's came to 3. Mr. P. and I divided Mr. V.'s, making mine 8, and Mr. P.'s 12, so we decided that it should be p. 12, line 8. We then washed the locked slates clean, locked them, Mr. Pinnock retaining the key. Mr. Davey placed his hand on the slates, and scratching was heard for a few seconds; on the slate being unlocked by Mr. Pinnock and handed to me, I found it was full of writing of a different character from that we had seen before; it consisted of an appeal to either Mr. Venner or myself, asking if one who was acquainted with electricity could fail to appreciate the difficulty of producing phenomena such as we were witnessing that evening; unfortunately the slate was washed before we had taken a copy. The writing in this case was not diagonal, but straight across the slate; it started about a quarter of an inch from the top of the slate, went right down to the bottom, then was continued round one side and finished up in the quarter of an inch left at the top of the slate, with two lines written *upside down*, and was signed with the initials T. P., I think. This

was interesting to us, but Mr. Davey was vexed we did not get the line out of the book written, and so, placing the slate under the table flap, he asked the reason; the word "muddle" was written, and we apprehended it was on account of Mr. Venner and myself both having chosen a book; we therefore thought it would be best for Mr. Pinnock, who knew Mr. Davey no better than I, to select another book.

[*f*] This he did. We washed the two slates, laid them face to face on the table, when the following words were written: "The difference in this respect." Mr. Pinnock now took down the book he had selected from the shelf, and handed it to me; I opened it at the 12th page and looked at the eighth line. I found the first two words completed a sentence; then came the five words above, and then two more to finish the line. I said the written words were right, but not complete. The slate was covered again, and three more words were written: "Shakespeare and Beaumont." On looking at the book I found Shakespeare was the last word in the line, the other two being in the next line. I said a word was still missed out. The slates were put together again, and two more words written. On looking at the book these turned out to be the two words terminating the last sentence. I said there was still the word missing, and this time the word "between" was written, making the sentence complete: "The difference in this respect between Shakespeare and Beaumont." I then asked for the last word in the line by itself, and this was written "Shakes," which was correct, as Shakespeare was half on one line and half on the other. The name of the book was *Lectures on Shakespeare, &c.*

[*g*] We next asked another question, and this time had the answer written on the *underneath* side of the *upper* slate instead of on the *upper* side of the *underneath* slate.

[*h*] Mr. Pinnock asked if we could not get the writing on a piece of paper instead of the slate. Mr. Davey said we might try, and thereupon tore a sheet of writing-paper into two, and pasted one half on to a slate by the four corners; he cut off a small piece of black lead from the end of a pencil, put it on the paper and covered the slate with another slate. Writing was heard at once, and we separated the slates and found the paper written over diagonally as in the case of the first slate. The paper was not, however, quite full, and it looked as if the slates were separated too soon, as the sentence was not finished. The writing was

evidently written with the point of the pencil.

[*i*] Mr. Davey was now very tired, but he offered to try one more experiment. A slate was raised on two glass blocks above the table, on top of the slate was placed a piece of crayon, and over the crayon was inverted a glass tumbler. Mr. Davey asked me what figure the crayon should draw. I said a triangle. We all joined hands and watched the crayon through the glass. After a few minutes, the crayon not having moved, Mr. Davey placed a slate under the table and asked if it would move, when the answer "No" was written, and we then finished our evening's experiments.

I have endeavoured in this report to merely give an account of what I saw, and not to give any attempt at an opinion as to the way in which the phenomena were produced; but this I may say, that it appears to me exceedingly improbable that electricity, as we at present understand it, was used. Everything occurred under full light and between the hours of 9 p.m. and 1.30 a.m.

E. MANVILLE.

DEAR MR. DAVEY,

I received your note yesterday just before leaving town. The writing always appeared on the upper side of the slate held against the table- flap; also the pencil was in every case, I noticed, at the end of the writing and decidedly worn, and in one or two cases, I recollect, on the last stroke. Will you kindly add this to my report. . . .   E. M.

11th December, 1886.
Oxford.

SITTING VIII.

My object in giving these séances has not been so much to "defy detection" as to enable some estimate to be formed concerning the possibilities of mal-observation and lapse of memory under certain peculiar conditions. Hitherto I have never refrained altogether from producing "phenomena" merely because I was afraid that the witness might discover my methods, although I have on several occasions given blank séances to persons who had already witnessed my phenomena, and whom I had no reason to fear. At the commencement of the sitting I saw that Mr. Dodds was an investigator who was justly entitled to a blank séance, and his account therefore is particularly interesting from the fact that notwithstanding his keenness, he failed to detect my real *modus operandi*.

Report of MR. J. M. DODDS.

12, Mitre Court Chambers, Temple, E.C.
19th December, 1886.

DEAR MR. HODGSON,

I now send you a report of our séance as I promised, for publication or not, just as you please.

On Mr. Davey's kind invitation I accompanied you last night to his house at Beckenham. There we dined, talking of telepathic and hypnopathic symptoms and similar subjects. I ought to say that my attitude was that of one totally sceptical regarding "spirits," very suspicious of trickery, and only in the faintest degree open to conviction that some quasi-explanation for the strange phenomena of which I had heard is to be found in the hypothesis of a new force or medium of transmission. I had never before (as I told Mr. Davey) attended at a séance. I had, however, some hearsay knowledge of his wonderful performances. But I did not know his point of view—i.e., whether he professed to act through "spirits" or otherwise; and although I tried to discover this, his answers and yours were so vague that I could not make sure. I inclined, when the séance began, to the opinion that Mr. Davey was a "believer," but was somewhat reassured as to his *bona fides* by his professed inability to imitate a simple conjuring trick which you showed us, and by his reminding me of some precautions which, in my inexperience, I was neglecting. Lastly, I am bound to say that although as Mr. Davey's guest I felt a little shy of showing my suspicions, I thought it all the more desirable to keep a close watch. This I was able to do as the room was well lighted throughout the evening.

I. The dinner-table was cleared and wheeled aside, and an uncovered ordinary Pembroke table was brought in. You and Mr. Davey left the room while I wrote a simple question in a small double slate belonging to Mr. Davey, which I carefully inspected, locked and kept within sight. The three of us then sat down at the Pembroke table, which we had examined. I unsealed a packet in which I had brought three new school slates; Mr. Davey chose one of them, which he and I, after making sure it was blank, held, in the manner to be described, under the corner of one of the extended table-flaps, with a small piece of pencil lying on its upper surface between slate and mahogany. The locked slate with the question inside was laid on the table—I had not let it pass out of my sight. Mr. Davey sat at a corner, his right hand and my left meeting on the under surface of the slate below the table-flap, while you, sitting opposite him, held his left and my right hand in yours. The problem as explained to me was: Given my question known to me alone; required an answer to be written upon my blank slate in position under the table, and to appear through some unexplained agency upon its upper surface where the chip of pencil lay; the answer either to give the information demanded, or at least to show knowledge of the question.

For several minutes we sat thus, either in silence or discussing psychical topics. Mr. Davey professed to expect no great success with me, and you reminded him of several séances which, after bad beginnings, had ended successfully. I asked some questions about the qualities required in the sitter, and, as before, received answers that did not enlighten me regarding Mr. Davey's standpoint, and therefore increased my vigilance.

No "phenomena" were forthcoming. At my request the slate (which, while underneath, I, of course, always pressed flat against the table) was now transferred to the top surface of the table, another was placed over it, and the pencil chip remained between, and Mr. Davey and I laid our hands upon the upper slate. No more success than before.

The conditions were subsequently twice varied. First my slate was restored to its original position under the table (said to be the usual one for preliminary manifestations), but with the stipulation on my part that I should keep it pressed against the flap with knee as well as hand; afterwards, deserting my slate altogether, we laid our hands upon the locked double slate containing the question,—but all in vain.

Finally, when more than an hour had passed, two of my slates, examined and found blank, were laid together, pencil between, and

The Possibilities of Mal-Observation, &c.

placed in position, like the single slate in the first effort, against the lower surface of the flap—our hands also remaining as at first. Very soon scratching was heard although I could detect no movement with my eyes, hand, or knee and, when the slates were brought to light, written upon the upper surface of the lower slate was the word "Yes." Now, as my question had been, "Where did I buy my slates?" I was not much struck by an answer that did not apply, and might have been written by some quite conceivable piece of jugglery; and my doubts were increased when I found upon the other side of the same slate, and therefore on the surface (such was my belief) where Mr. Davey's hand had rested, the word "Wait." I was, therefore, very little impressed by this result; and indeed, rather to my surprise, neither Mr. Davey nor yourself seemed to expect me to draw any conclusion from it.[1]

A subsequent experiment—in which I repeated my first question, carefully expunged from the double slate and still unknown, except to myself—upon one of the open slates, came to an abortive ending through Mr. Davey's catching sight[2] of what I had written.

II. After an interval, Mr. Davey, who acknowledged that he was not in a good frame of mind for "manifestations," was induced by you to try the "book" experiment. This was explained to me to consist in my mentally choosing from the books, which, to the number of, I should guess, about 700, [1000] lined the room, any one with a clear title; I was then to take twice over a handful of fragments of slate-pencil from a box on the table, privately count each handful before replacing the fragments, and keep the results to myself the first result was to represent the number of a page of the book chosen, the second the number of a line on that page; Mr. Davey, yourself, and I were to lay our hands upon his double slate, laid upon the surface of the table after being examined, found blank, and locked with pencil-chip inside by me; I was to concentrate my thoughts upon the book and numbers and Mr. Davey was to try to discover (by some mode of thought-transference, I inferred) book, page, and line: the pencil locked inside the slate was then to write some words quoted from the place thought of!

The preliminary programme was carried out, and I may say that while choosing the book I took care to walk right round the room and not to let my eyes linger on any one spot. Thus the problem was: Given a book, page and line known only to me and recorded nowhere—

required to be written in a blank locked slate lying under our hands the corresponding quotation, which, be it observed, was unknown even to me, for of course I had not touched, much less opened the book.

This appeared impossible by any amount of jugglery, and I could scarcely take the attempt seriously. We sat down, however, and laid our six hands upon the slate. I concentrated my mind with the utmost intensity at my command upon book, name, and numbers, and soon Mr. Davey appeared to labour under some excitement, and, to my disgust, began (with an explanation that it was the custom) to invoke some unseen agents in an appealing tone. Presently, to my relief, he desisted, and the attempt was given up as a failure. Mr. Davey said he could not decide between two books.

After a short rest it was suggested that I should name the book, and that the experiment should be resumed in a modified form. The problem was now: Given a certain book, viz., *Taine* on *Intelligence*[3]; required to be written in a blank locked slate, lying under our hands, a quotation unknown to anyone present, taken from a page and line known only to myself. The book, of course, remained untouched on the shelf. We sat as before with the slate under our hands and eyes. I concentrated my thoughts. Mr. Davey soon appeared to reach a high pitch of exaltation; his arms and body became subject to a violent "*frisonnement.*" He again appealed to his ghostly helpers, and on this occasion his efforts were rewarded, for, in a few minutes, to my utter amazement—Mr. Davey's hands and your own being well in sight and unemployed—I heard sounds of writing within the slate which continued for half a minute or more. On unlocking the slate I found, legibly written, a quotation, almost, but not quite, verbally correct, from page 15 of Taine's book, beginning at the eighth line. Some "clear-obscure" remarks, which I at once interpreted as relating to a friend of mine, followed.

I had thought of the eighth line of the 28th page. The correspondence was, therefore, not exact, the line only being correct. What struck me, however, was not the coincidence of the quotation, nor the gibberish about my friend, which hinted information easily ascertainable by anyone who, like Mr. Davey, had met him—it was the occurrence of what the evidence of my senses told me was writing by a piece of inanimate pencil inside a locked slate, with no conceivable means of explanation! For a moment I confess I was completely staggered; my notions of causation were turned topsy-turvy; visions of "magnetic force" and "occult action" danced before my brain. Then came the reaction; but instead of accusing my senses of perjury, I illustrated human nature by telling you in plain English (during a momentary absence on Mr. Davey's part) what opinion I had formed of him. I regret to think I used the word "humbug"; none could be less applicable!

I had not just then much desire to continue the séance; but you seemed to desire it, and as I recovered from my bewilderment, one or two slight circumstances—one of them Mr. Davey's half acquiescence in a suggestion that he should try to obtain writing *without any pencil sandwiched in the slates*—occurred to me as confirmatory of my notion that he had been slate-shuffling in some very clever way. So I asked him point-blank, as you remember, what was his theory; he answered that he does not so far profess any theory, but merely undertakes a close imitation of the phenomena attributed by believers to spirits. I had not quite realised this before, and was now for the first time able to appreciate Mr. Davey's standpoint—though no less in the dark as to his method. We seemed, as you afterwards remarked, to "have an understanding" from this time; and with my good temper I recovered my vigilance.

III. The last experiment consisted in my writing a question as at first in the locked slate, to be answered by writing produced between two of my plain school slates by chips of pencil; the slates having, of course, been examined and found blank as usual. The slates were laid upon the table-top, and except that I had unintentionally changed my seat to your former one opposite Mr. Davey—you taking mine in exchange—the conditions were as before. We laid our hands on the upper slate, but after several efforts no result was obtained. We were proceeding to make another trial, and Mr. Davey, in the act of displaying the slates to show that they were still blank, made a remark to you which had the effect of causing me to look at you; just then, more by accident than design, I noticed that before replacing the upper slate upon the lower one he reversed its position. Seizing it at once, I found one of its sides—that which would have been underneath—covered with an inscription which I certainly had not seen or heard written, and which in my delight I forgot to read. Evidently the next effort would have been the success of the evening!

The game was up; at least you and Mr. Davey chose to think so, for you at once let me into the secret of the great Psychical Plant. I don't profess to understand Mr. Davey's *modus operandi*; but of this I am certain, that I have to thank you for an introduction, not to a world where the rules of nature are superseded, but to a most surprising exhibition of sleight of hand.—Yours very truly,

JAMES M. DODDS.

SITTING IX.

I had never seen either of the writers of the following accounts until the day of the sitting, but I understand that they had already learnt beforehand that what they were about to witness was unquestionably due to conjuring.

1. *Report of* MR. A. B. T.

Monday,
Grosvenor-place, S. W.

DEAR MR. DAVEY,—

I am just writing a small account of what I saw you do last Thursday night (as you requested), at Mr. T. B.'s house, when I had the pleasure of meeting you, and witnessing your wonderful feats of slate-writing. The following is as near as possible what took place, by memory: We, a party of five (exclusive of yourself), were sitting in the drawing-room, round a plain deal table with flaps, which had been brought from the kitchen. You provided three ordinary slates and a small handsome lock-up slate with a lock and key. There was also a sponge, cloth, and glass of water on the table, with which I cleaned the slates. The first thing you did was to give me the small lock-slate to examine, and having assured myself that no trickery existed in it, I cleaned it and placed a small piece of coloured pencil on it, locked it up, and put it in my pocket. The key I placed in my waistcoat pocket.

[a] You then took a point of pencil and laid it on the table, over which you placed one of the common slates which I am positive I had

The Possibilities of Mal-Observation, &c.

thoroughly sponged and wiped. We joined our hands, and you and I placed ours firmly on the slate. You asked your spirit-friend "Joey," if he could give us any help, and very soon after an extraordinary sound of scratching was heard under the slate. Upon raising it, the following appeared in large bold letters right across it: "Allright; here we are again. Hurrah! —Joey." This was very satisfactory, and "Joey" worked very hard to answer us afterwards.

[c] For you next held a slate with a small piece of pencil upon it under the flap of the table, and wished a question to be asked. In reply to mine, as to when my train would arrive at Victoria, the reply came very soon, "*Wait.*"

[d] This was not considered a sufficient answer from the [spirit?] world, so you quickly rubbed the slate, and immediately held it again under the flap of the table. We waited some time, and then got some writing as before, "*No chalks,*" and on looking at the slate I saw you had forgotten to place the chalk upon it. It was expecting too much of "Joey" to write without a chalk.

[e] You then took two slates, which I once more wiped clean, and placed them one upon the other on the table, with a tiny piece of pencil between them. There was a very short wait, and then the sound of quick writing was heard. This lasted for nearly a minute. Upon raising the slate, upon the top one was written as follows:—

"Dear Friends,—It is not so much the agency question we would have you set your minds upon, as it is the mere fact that the phenomena take place under conditions which upon every reasonable mind preclude the *possibility* (?) by known rational means. You may rest assured we shall do all in our power to co-operate with you this evening; we must, however, ask you to have patience, as we can't carry out any tests or answer any question until we have become more *en rapport* with one another. Rest assured and we will do our best, and remember Der Teufel is zu zwart nit als hig wel geschildered. —*Joey.*"

[f] We next experimented with the slate which I took from my pocket. You asked me to choose a book from the bookcase, and one of the party to think of the page and line. I went to the bookcase, and could not make up my mind between three or four, and finally took Virgil's *Aeneid*. The slate was placed on the table, and "Joey" was again asked to write a passage from the selected book (in any coloured chalk

I liked). Again the scratching of the pencil was heard upon the slate. When it was opened, the piece of chalk was nearly worn away, and rested upon the last word of the following: "We should prefer that when you experiment for tests, such as the one you now propose, that you should form a smaller circle, and devote yourself exclusively to this one form of phenomena, and although it is not impossible we may succeed tonight, yet we are greatly hampered by the co-operation of too many minds. We have no objection to try the tumbler, although we don't guarantee a —— ERNEST."

This was the last experiment tried, as time drew on, and I wanted to catch my train back to town. As to that part of the test relating to the passage chosen from the book, it failed.

I was very sorry to see you in such weak health, and the excitement under which you laboured showed plainly that the mental strain upon you must have been great. I noticed upon every occasion of the writing appearing you trembled and shuddered as if under great nervous pressure, but why this should be, if, as you say, these manifestations are only the result of trickery and conjuring, I do not know. At all events, you have mystified me entirely. I do not believe in spiritual manifestation in the least, but how you manage to bamboozle so many people I can't make out.—Believe me, Mr. Davey, sincerely yours, A. B. T.

2. *Report of* MISS M. T. B.

Whilst staying with my uncle at Beckenham I had the good fortune to meet Mr. S. Davey, and to witness some of his interesting manifestations. I had heard of his wonderful powers, and was therefore very pleased when I learnt that he had accepted my uncle's invitation, and had volunteered, after dinner, to show us some of his experiments.

We were five in number, and were seated round an ordinary deal table, which had previously been carefully examined so as to preclude any possibility of trickery.

Before commencing the séance, Mr. Davey produced a book-slate, carefully cleaned it, and gave it to one of our friends, asking him to place a small piece of pencil in it, lock it up, and put it with the key into his pocket. There it remained until later on in the evening it was required for use.

[a] Our first experiment was with an ordinary school slate. Mr.

Davey placed a piece of chalk on the table, sponged and wiped this slate perfectly clean, and placed it upon the chalk. We all joined hands, Mr. Davey resting his upon the slate. After a few minutes a faint scratching was heard, and on being examined, the slate was found to have written upon it in good bold characters, "Hurrah, here we are again, Joey." This seemed to me most wonderful, as all the time Mr. Davey's hands were visible.

[b] Next, Mr. Davey placed a piece of chalk upon a slate, and put the slate under the table, supporting it with his right hand. After listening for some little time, we again distinctly heard a faint scratching, and in answer to the question whether we should have any manifestations that night, we found the answer, "Yes."

[c] Again a question was asked as to the time of the departure of the last train to London Bridge, and in reply we found "Wait" written upon the slate.

[e] Mr. Davey then volunteered to produce writing in two differently-coloured chalks, and the two selected (blue and white) were placed between two ordinary slates. Again we joined hands, Mr. Davey resting his, as before, upon the slates. After waiting for some little while (in this case longer than previously) the scratching was heard, and upon examining the slate, it was found to be covered in writing, half being in blue chalk, the other half in white.

[f] Mr. Davey now asked for the book-slate, and requested one of our friends to think of two numbers, then to select a book from the bookcase, taking care to keep the title of the book well in his mind. Mr. Davey proposed to produce the quotation from the chosen page and line of the book. Unfortunately, with this trick there was a little mistake, as our friend glanced at several books before settling which he would finally choose, and the quotation consequently was not produced. Instead of the quotation, some advice was found to be written. I do not doubt that this failure was caused by the want of concentration of mind upon the chosen book. With this trick I was particularly struck with the fact that there was a visible difference in the size of the chalk when placed upon the slate and when it was examined after the writing had been produced, and also the remainder of the chalk was discovered at the end of the last word written; both these facts seemed to me to prove that the writing was produced by the chalk alone, and by no other secret agency.

SITTING X.

The following report is by a Japanese gentleman whom I had met once previously, and who attributed sundry phenomena of "mediums," which had been discussed, but which he had not personally witnessed, to the action of some new unrecognised force.

Report of MARQUIS Y. A. T.

[a] On January 24th, 1887, I had the pleasure of seeing Mr. Davey's slate-writing performances in his private room. He first removed a table, which was in the corner, into the middle of the room, and brought two common slates, one double-slate which can be locked, a sponge, a box of chalks, and a glass of ordinary water. Then he told me to examine and sponge and wipe them thoroughly; so I did, and put first the common slate, a piece of chalk being under it, and put our hands on it; a faint sound of scratching was heard. When the sound had ceased, Mr. Davey turned it out as follows: (No. 1) was written on it.

No. 1.

"Japanese is very difficult language to write, but we will do our best. We are sorry not to see Baron this evening, please give him our kind regards.—JOEY."

[b] Next I locked the double-slate (small piece of chalk was put in it), and laid before him, and we put our hands on it while I was holding the key in my hand. As soon as the scratching sound had ceased I unlocked it and found such words as follow:—

No. 2.

"A student like yourself will easily understand the importance of this writing, which, as far as the senses are able to judge, would appear to be of a supernatural character. This, however, is not the case. In order to prove to you that we are above ordinary conditions of conjurers, and also demonstrate the absurdity of writing being produced by chemical action we are willing to carry out any test you may suggest which would serve to dispel such a theory from those who have not witnessed these performances."

[c] Again I put another common slate on the table and we put our hands on it. In turning it we found a Japanese but really Chinese character was written.

No. 3.

A Chinese letter 散. This letter is also used in Japan because the Japanese are using the Chinese characters. In Japan this letter is used as a verb, and means "to be scattered" or "to be dispersed." It is pronounced "Chinu" in Japan.

[d] Once more I locked the double-slate (this time I put white and blue pieces of chalk in it), and put the key in my pocket and even sealed it myself. In opening it I found a letter in Japanese character was written and also an English as follows:—

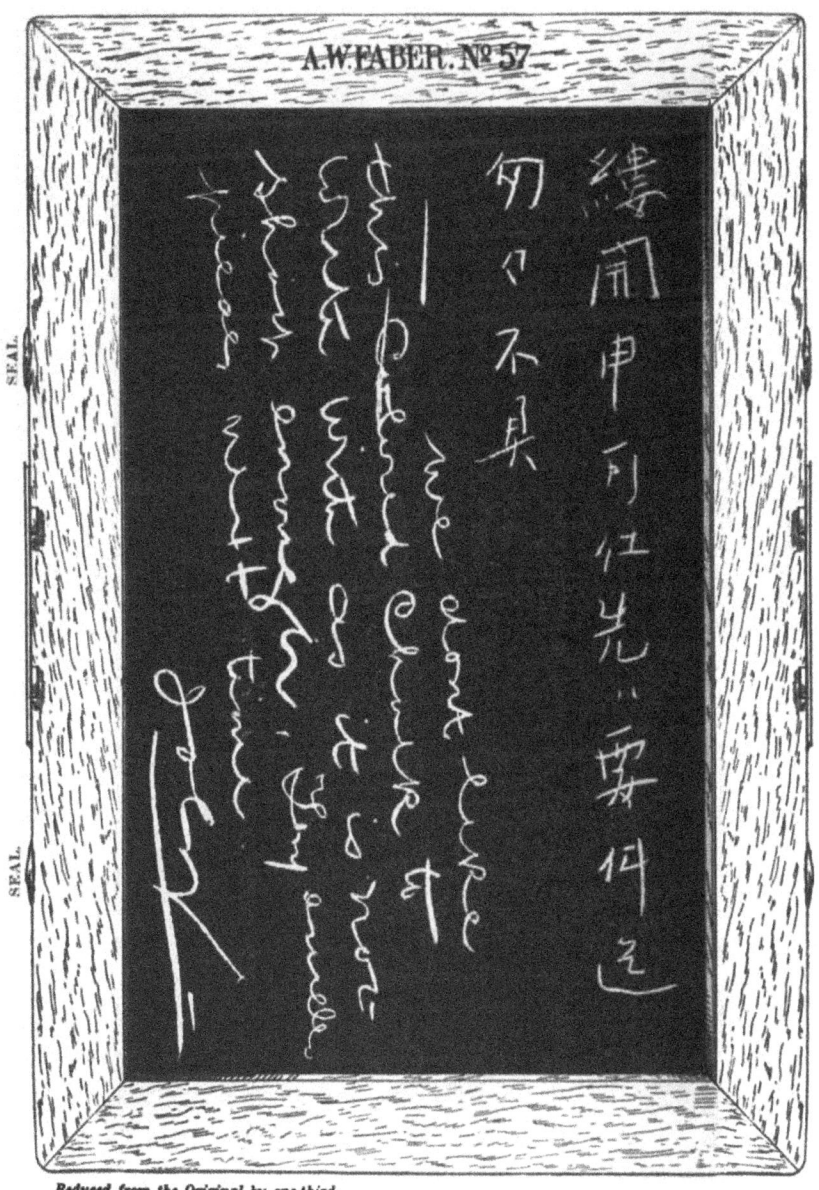

The Possibilities of Mal-Observation, &c.

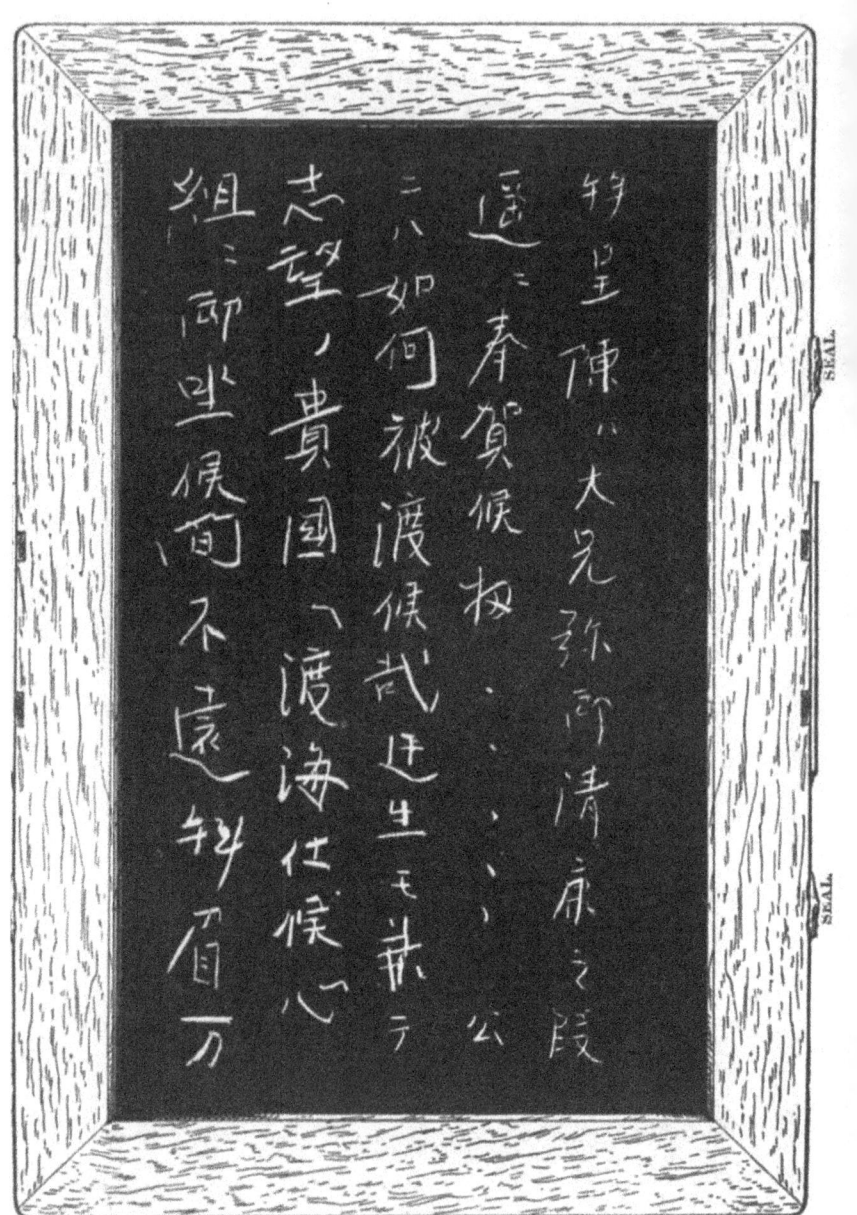

A.

Hai tei shikareba taikei go-seikô no dan haruka ni gasi tatematsuri soro sate wa * * * kô ni wa ikaga wataraserare soro ya usei mo kanete shibô no ki-koku ye tokai tsukamatsuri soro kokoro-gumi ni goza soro aida tôkaraz haibi banru kaishin tsukamatsuru beku madzu wa yôken made sôsô fugu (in Japanese character, with white chalk).

Japanese, when translated, relates to my friend Prince * * * in Japan, and it also means it is pleased to give me writing in Japanese.

B.

We don't like this coloured chalk to write it as it is not short enough. Try smaller piece next time.—JOEY.

This performance was more than wonderful, and I could not see how anything could have been written. But I hope I will have an opportunity to learn about this by reading Mr. Davey's excellent work on this subject. I had asked for writing to come in blue colour, and writing in blue English colour is an excuse for not doing so, I suppose.   Y. A. T.

SITTING XI.

Mr. H. W. S., the writer of the following record, was a comparative stranger to me, as I had met him on only one previous occasion. He requested me to give him a séance, as he had heard of my performances from some of his friends who are well-known to me. Previous to the sitting I informed him, as I have also informed many others, that my phenomena were not to be accounted for by the "Spiritualistic" theory.

Report of Mr. H. W. S.

February 11th, 1887.

After the very interesting scientific phenomena to which I was an eyewitness last night, it gives me much pleasure to detail the various astonishing feats displayed by Mr. Davey [*v.* Clifford.]

The apartment in which I was received was a well-stocked library,

The Possibilities of Mal-Observation, &c.

and the furniture, including the table at which we sat, was of the ordinary make and style, with none of the intricacies so necessary to the everyday conjurer; and I am convinced that the furniture of the room and its general surroundings played no part whatever in the accomplishment of the facts which I am going to narrate.

Having produced a small book-slate, Mr. Davey asked me to examine it, and to satisfy myself as to its simplicity of construction, &c. I did so; the slate was composed of two ordinary pieces of slate, about six by four inches, mounted in ebony covers hinged on one side with two strong plated hinges, and closed in front, beyond the question of a doubt, with a Chatwood's patent lock.

With the exception of a small escutcheon, bearing the initials of the donor, the slate was plain and substantial, and bore the strictest inspection, so as to entirely preclude the idea of chemicals or any other similar agent being used to it.

[a] After I had finished examining the slate, Mr. Davey asked me to write in the slate any question I liked while he was absent from the room. Picking up apiece of grey crayon, I wrote the following question: "What is the specific gravity of platinum?" and then having locked the slate and retained the key, I placed the former on the table and the latter in my pocket.

After the lapse of a few minutes I heard a distinct sound as of writing, and on being requested to unlock the slate I there discovered to my great surprise the answer of my question: "We don't know the specific gravity, JOEY." The pencil with which it was written was a little piece which we had enclosed, and which would just rattle between the sides of the folded slate.

Having had my hands on the slate above the table, I can certify that the slate was not touched or tampered with during the time the writing was going on.

[b] Next; having taken an ordinary scholar's slate and placed a fragment of red crayon upon it, Mr. Davey placed it under the flap of the table. I held one side with my hand as before. I then heard the same sound as previously, and when the slate was placed on the table I found the following short address distinctly written: "Dear Mr. S ——, — The substitution dodge is good; the chemical is better, but you see by the writing the spirits know a trick worth two of that. This medium

is honest, and I am the only true JOEY." The writing was in red crayon, and was in regular parallel straight lines.

[*c*] Then, again, Mr. Davey requested me to place a small fragment of slate-pencil in the lock slate, which latter had been previously cleansed with sponge by *me*. Respecting the method of closing the slate, &c., everything was done as in the *first* instance; the slate was locked, and I retained the key.

As soon as the sound of writing was over, I picked the slate from off the table, where it had been lying right under my eyes, unlocked it, and read as follows: "We are very pleased to be able to give you this writing under these conditions, because with your special knowledge upon the subject you can negative the theory of antecedent preparation of this slate as advanced by certain wiseacres to explain the mystery.—'JOEY.'" The fact that the pencil when removed from the interior of the slate had diminished in size and showed distinct traces of friction, convinces me that it was the pencil and nothing else which produced the caligraphy. If the particles taken from the pencil by friction did not go on the surface of the slate, where could they go?

[*d*] Lastly, as requested by Mr. Davey, I took a coin from my pocket without looking at it, placed it in an envelope and sealed it up. I am certain that neither Mr. Davey nor myself knew anything about the coin. I then placed it in the book-slate together with a piece of pencil, closed it as previously and deposited it on the table; and having placed my hands with those of Mr. Davey on the upper surface of the slate, waited a short time. I then unlocked the slate as requested, and to my intense amazement. I found the date of the coin written, by the side of the envelope containing it.

The seal and envelope (which I have now) remained intact.

This last feat astonished me more than the others, so utterly impossible and abnormal did it appear to me. I may also mention that everything which was used, including the cloth and sponge with which the slates were cleansed, were eagerly and thoroughly scrutinised by me, and I failed to detect anything in the shape of mechanism of any kind. Were I sceptically inclined towards Spiritualism, I should have attributed the feats I witnessed to it, but I am convinced from the bona fide manner in which Mr. Davey proceeded to perform his mysterious writing, Spiritualism plays no part in it whatever. Were I asked to account for the method by which the writing was done, or rather to advance any theory based upon which it would be possible to produce such phenomena, I should suggest a powerful magnetic force used in a double manner, i.e., 1st. the force of attraction, and 2nd. that of repulsion.[37]

But Mr. Davey has by great perseverance and study cultivated his scientific secret to such an extent that were it magnetism, electricity, pneumatics, or anything else, it would baffle the most accomplished in any of those branches of science to form even an approximate idea of his *modus operandi*.

SITTING XII.

Mr. Henry Hayman was introduced to me by Mr. H. W. S., the writer of the preceding report. In connection with this sitting I may observe that not only is it commonly the case that witnesses believe themselves to have taken precautions which they did not take, but that they also frequently omit to record precautions which they did take.

In the present instance, for example, Mr. Hayman suggested during one experiment that it might be said that I produced the writing by means of mechanism connected with my knees; I thereupon desired one of the sitters to look under the table during the continuance of the experiment; Mr. H. W. S. proceeded to do so, but was unable to discover anything suspicious.

Report of Mr. Henry Hayman.

February 16th, 1887.

[a] Mr. H. W. S. and myself visited Mr. S. J. Davey last evening, and he gave us a séance of his slate-writing. He first got a plain table, without any secret contrivances, let me examine it, which I did thoroughly, then brought two ordinary school slates, and asked me to wash

---

[37] Compare the theory framed by Dr. E. von Hartmann to account for slate-writing phenomena, in *Der Spiritismus*. See C. C. M.'s translation, *Spiritism*, pp. 45-48.—S.J.D.

them with a sponge and water, so that if there was any writing on, it might be washed off. He next placed one flat on the table, asked me to take two small pieces of chalk, and put them on the slate, which I did, a blue and a red piece; then he put the other slate on the top, and we joined hands and pressed them on the top of the slates. After a minute or two I distinctly heard writing; when it had ceased he asked me to lift it up. When I did so the bottom one was covered with writing in the following words (half in blue and half in red chalk): "You will please pardon us friends if we do not enter with you tonight into experiments of a very minute nature, the grand fact of this writing should be sufficient, and we do not care to cloak the wonderment thereof by descending to underhand coin tricks, and such like phenomena, which however startling to some would fail upon those who like yourself are acquainted with conjuring possibilities. Good-bye."

[*b*] The next thing he showed me was a slate which locked up with a patent lever lock. After I had washed the slate, he asked me to write down on the inside any question I liked, then put a piece of chalk in, lock it up, and put the key in my pocket. The question I asked was, "What kind of weather shall we have tomorrow?" He was out of the room while I wrote it down, and it was locked up by the time he came back; he then placed it on the table, the gas being alight at the time, we joined hands and put them on the top of the slate. After a little I again heard writing, and when I opened it there was the answer, in red chalk, each side of the slate: "Ask the clerk of the weather." It had been written with the piece of chalk I had put in. I am quite certain the slate had not been opened after I had locked it up.   HENRY HAYMAN.

P.S.—I may add that I watched Mr. S. J. Davey very closely, but I could see no possible means by which any of his slate-writing could be done by ordinary known means.   H. H.

Statement of MR. H. W. S.

February 16th.

It is with still greater astonishment that I, in company with my friend, Mr. H., witnessed last night a séance of, if possible, greater anomalies than the previous one. Whilst confirming all the details given by my friend, Mr. H., I should like to draw attention to the fact that there were two witnesses in the matter, whose evidence is undoubtedly more reliable than that of one person. Moreover, it is far more difficult to perform in their presence, as the observation is more intense.

[H. W. S.]

SITTING XIII.

Mr. Stanley W. Jones, the writer of the following report, was inclined to account for the phenomena which he witnessed, and for analogous "mediumistic" phenomena, by some theory involving the action of magnetism.

Report of MR. STANLEY W. JONES.

I called on Mr. S. J. Davey, by appointment, on the evening of Friday, March 25th, 1887, bringing with me two perfectly new slates, which I had purchased that day and privately marked. I was shown into a well-furnished study, but without any of the usual paraphernalia of the conjurer about it. An ordinary dark wooden table with two flaps was brought forward, which I carefully examined. I pulled out the single drawer, finding it full of papers, and nothing suspicious or mysterious about it in any way.

[*a*] In answer to my request, Mr. Davey took an ordinary slate of his own, which I carefully sponged and wiped. He placed the same under the leaf of the table, I putting between the surface of the slate and the wood of table a piece of crayon. I supported one side with my left hand, he the other with his right. His left hand and my right were clasped. He asked me to propound some question. I accordingly submitted the following interrogatory: "On what day of the week does the 1st April fall?" After a certain interval he said, "Let us examine the slate." It was raised, but apparently presented an unchanged appearance. After replacing it as before, in a minute or two I heard a very faint scratching, and on looking at the surface the following fairly legible scrawl appeared: "I don't know."

[*b*] Mr. Davey now placed in my hands a small ebony backed slate, with Chatwood's lock; bidding me, while he retired from the room,

write any question therein. I inscribed, "What was the exact date of my birth?" placing the key in my pocket. I put, however, a new piece of crayon inside. The slate was now laid on the table and we placed our hands upon it. After a short time the scratching was again heard, and on unlocking the slate the following perfectly legible answer appeared: "I do not know your birthday." The facets of the crayon, which were perfectly unbroken on being put in, I found worn away, and resting on the last y.

[c] I now took the two new slates which I had purchased, and which had never for a moment passed out of my possession, I even taking the precaution of sitting on them during the foregoing proceedings. I placed a piece of red crayon therein, and screwed them down top and bottom so tightly that by no possibility could even the thin edge of a penknife be introduced. I then corded the slates twice across and across, sealing them in two places with red and blue wax (for of course any attempt to remove the seals by heat would cause the colours to fuse, and thus immediately detect the artifice), stamping them with my own private signet. Mr. Davey placed the slates under the table, and requested me to name some word I would like written. I stipulated for "April." After a few minutes, during which I most carefully watched him, he returned them, and after 10 minutes' work, so tightly were they closed, I found exactly what I had desired.

[d] He next took his own slate again, laying it upon the table, I having previously placed a piece of crayon therein, and pocketing the key after locking it. After a slight interval, a distinct and continuous scratching as of regular writing was heard, which lasted exactly 25 seconds. On unlocking the slate, the following message in a clear running hand appeared: "Dear Sir,—We regret to think that in this enlightened age there are still to be   und investigators in the realm of Trixography whose minds do not aspire beyond the pencil thimble, and 'half-a-crown' trick slate. We are therefore pleased you should impose tests upon our powers, as we are anxious to thoroughly satisfy you of our medium's honesty. Good-bye."

This concluded a most interesting, successful, and perfectly satisfactory psychographic séance.

The precautions I took entirely preclude any idea of trickery or conjuring. It is also a noticeable point that whereas the last "message" was concluded in 25 seconds, it takes at least five minutes to transcribe. I was extremely careful in doing my utmost to detect any artifices of Mr. Davey, but must own that not only was I most completely baffled, but everything seemed perfectly open and above board, the entire proceedings being transacted in the full glare of two powerful gas jets. I left very much impressed with Mr. Davey's marvellous powers and the phenomena I had witnessed.

STANLEY W. JONES.

[c] P.S.— *Nota Bene.*—After perusal of above, considering that the expression, "I found exactly what I desired," might be liable to a possible misconstruction, I think it better to add that I state in the most unequivocal, explicit, and emphatic manner, that after Mr. Davey had returned me my two slates, secured as above described, and which I *most* carefully and minutely examined to detect any signs of tampering, finding however my seals intact and the cording and screws in exactly the same condition as when they left my possession a few moments before, and that the word "April," which I had asked for, was legibly written with the crayon, on one of the inside surfaces. Whether the top or bottom I did not observe. The apparently impossible having thus been solved as I hereby testify.

STANLEY W. JONES.

SITTING XIV.

The following account is by a gentleman who has had considerable opportunities of observing and taking part in experimental inquiry. I shall speak of him as Dr. Q. He was aware that my performances were conjuring, and had read my paper in the Journal of the S.P.R. for January, 1887. He had had sittings with "slate-writing" mediums, and knew precisely the object of my investigation, although he did not know the methods which I employed. His account is remarkably good, and I quote it chiefly to show the rapidity with which slates fastened together as described can be manipulated.

Experimental Investigation

Report of Dr. Q.

[*a*] I have witnessed this afternoon, about three hours ago, some remarkable phenomena, when sitting with Mr. Davey, which I think worthy of a brief record without comment. The séance was in a friend's room, in the early afternoon, in thoroughly good daylight. An adjoining room, entirely separate from the experimenting room, gave me the opportunity of making my preparations out of Mr. Davey's sight. I took three common schoolroom slates and examined them carefully by myself, testing the security of the joints of the wood frames, the solidity of the piece of slate enclosed, and its freedom from marks of chalk. I had brought a strong lens and was thereby able to mark for identification each slate, A, B, and C, with marks not visible to my own eyes without a lens. I then laid B on C and tied them tightly round with strong string whose elasticity was very slight. The knots I sealed and marked the seals. I had left a small piece of whitish chalk between the slates. I brought them in this condition back to the experimenting room where Mr. Davey was, and we seated ourselves at the corner of a wooden table without tablecloth. The only drawer in the table which I could find had been removed. I gave the slates to Mr. Davey, who was sitting on my left. He took them and held them in his two hands under the table for a minute or more,[(1)] then gave me his left hand above the table, and I held it in my left hand. My knees under the table were separated by a leg of the table from free movement towards Mr. Davey. After two or three minutes, during which there had been some conversation, he asked me to suggest some word which should be written between the slates. I said at first Vladivostok, but as he thought this inappropriate I changed it for Hong Kong. I heard scratching sounds, but observed less movement of Mr. Davey's right elbow than there had been before. After a minute or two he reproduced the slates and asked me to examine them and see what had been written inside. I found no appreciable displacement of the string, no change in the seals, and the marks of identification were clear. I cut the string, therefore, and opened the slates. On one I found rather indistinct scratches in an apparently continuous line, of which one-half bore a considerable resemblance to a badly-written Hong; the other half was of about equal length, but not decipherable.

[*b*] I retired to the preparing-room to make ready for another ex-

periment. I drew on the internal surface of B five vertical lines, and on the internal surface of C five vertical and five horizontal lines with the fragment of whitish chalk and tied and sealed them as before. This time I thought that the apposition of the edges was more complete and secure than before. I brought them to Mr. Davey, and after they had been handled as before, I asked that the word Irishman should be written, but there was no success; I did not find any appreciable alteration on the inside or outside of the slates. My left knee had been under the apparent position of Mr. Davey's right hand.

[c] In the third experiment the same slates were used; they were prepared in Mr. Davey's presence. They were tied and the knots sealed and also two screw nails were driven through the wooden frames of both slates, one at what I may call the south-east corner and the other at the middle of the north end. This made firm and satisfactory apposition of the frames in the neighbourhood of the screws. The screws were not sealed, but the position of the fine broken woody fibres round them was noted. I did not succeed in making the tied string quite as tight as in the previous experiment. I noticed no difference in the manipulation of the slates from the first two experiments. I made no attempt to touch Mr. Davey's hand with my left knee. I asked that 77 should be written, and after a few seconds explained myself by saying that they should be written out in full, not in numbers.

The slates were brought above the table in less than a minute after this. I could find no change externally, and written inside I found the figures 77, and also a fairly well-formed "Sev," followed after an indistinct interval by signs resembling "ty," and these also by some almost indecipherable scratches.

[d] For a fourth experiment I prepared the same slates in the adjoining room. A fresh screw was fixed in the previous hole near the south-east corner, and another was passed through both frames a little below the north-west corner. The exact position of the heads of the screws was noticed, but they were not sealed. I carefully examined the piece of chalk left between the slates with my lens, and found two smooth rubbed facets on it. These and the other rough surfaces of the irregular polygonal mass I marked with very finely cut lines. The slates were tied with thinner and more yielding string, and the knots sealed. A rather fragile drop of sealing wax was placed on

the junction of the crossed strings over the middle of the slate. The chief object of the experiment was to test the rapidity with which the phenomena occurred when Mr. Davey was not under my observation. I was to stand with my back to Mr. Davey, to hold out the slates to him behind my back, and to order some word which was to be written between them behind me out of my sight, and the slates returned to my hand behind my back as soon as possible. I was to note the time spent on this. On a first attempt I held out the slates behind my back saying, "commandment." Mr. Davey took them, asking at the same time what was the word required. I saw that I had not made myself plain, turned round and took back the slates within one or two seconds, and found no change externally except that the fragile drop of sealing wax on the crossed strings was broken. I took the slates back again to the preparing-room, and put on another similar drop. On a second trial I again said, "commandment," putting the slates behind me with my left hand, holding my watch in my right, and keeping my back to Mr. Davey. They were returned to my hand in 30 or 31 seconds. Some sounds which might be described as shuffling and scratching had been heard by me. I laid them on the table, and on examination found no change externally; then cut the strings and found on the inner surface "commandment" written more distinctly than the previous words had been and near the edge[2] of the slate. Whilst I was looking at the word Mr. Davey touched and to some extent broke the piece of chalk. In examining it afterwards with a lens I found only two smooth rubbed facets, each of which was still marked with the finely cut line I had made on it over some only of the remaining rough surfaces could I find the fine lines that I had previously made over all. I made no accurate comparison between the marks I could produce on the slate by the whitish chalk and those forming the words or parts of words. In no case was the piece of chalk large enough to touch both internal surfaces at the same time.

March 25th, 1887.

SITTINGS XV. AND XVI.

Miss Symons was introduced to me by Mrs. Sidgwick on the day of our first sitting. I gave to her and Mrs. Sidgwick three sittings in all, of the second of which, a dark séance, she wrote no account. As I have already stated (p. 92, note), I shall not quote the accounts written by Mrs. Sidgwick, as I had told her a good deal about my tricks beforehand, and she knew that my performances were conjuring. Notwithstanding this, however, Mrs. Sidgwick was unable to explain some of my phenomena. At my request, she has furnished some notes concerning the reports of Miss Symons, and she also makes the following statement: "I did not communicate what I knew about Mr. Davey to Miss Symons till after the second séance here recorded, as my object in taking her to the séances was to obtain an account of what she witnessed, written as nearly as possible in the same state of mind as when she wrote certain accounts (printed in the *Journal* of the S.P.R. for June, 1886) of Mr. Eglinton. I therefore merely represented Mr. Davey as a person through whom remarkable phenomena occurred, which I wanted to have investigated. Mr. Davey himself seemed to me (as I have recorded in my notebook) to talk very openly to Miss Symons. He seemed to tell her almost as much as he had told me about his tricks and those of other 'mediums,' but it was doubtless mixed up in a mystifying way. Unfortunately, Miss Symons had great confidence in my care as an investigator, and, without revealing the actual situation, I could not succeed in making her feel herself dependent entirely on her own observation. It is due to her to state this, as, had she left me out of account she believes she would have used more precautions than she did, and she considers that she was more careful in her investigation of Mr. Eglinton than in that of Mr. Davey."

These sittings were the earliest I gave which were recorded in detail; my experience then was comparatively limited, and I have since become much more practised in certain methods, and have also acquired the knowledge of new ones.

Experimental Investigation

SITTING XV.

Report of Miss Symons.

Slate-writing séance, November 16th, 1885, at 14, Dean's-yard, with Mrs. Sidgwick, and a medium whom I will designate as Mr. A. [changed throughout to D., see p. 86]. Our sitting commenced at 7.45 p.m. We took our places round a deal table in the following order: —The medium Mr. D. at one corner, next him Mrs. Sidgwick, and I opposite.

Neither Mrs. Sidgwick nor I had brought any slates, and we were, therefore, obliged to use those brought by Mr. D. We sat in a good light, a lamp and several candles were burning in different parts of the room. We first washed the slates ourselves with water brought us by Mr. Podmore, so that there was no question of its containing any admixture of chemicals, by which means writing might be produced, as has sometimes been suggested to me; the table, too, was above suspicion, having just been bought by Mr. Podmore for this particular séance. After each one of us had separately washed and dried the slates, one was marked by Mrs. Sidgwick, a piece of pencil was placed on it, and it was held by Mr. D. under the table, who warned us to watch him very carefully, as he gave no promise not to cheat, did we give him the faintest opportunity for so doing, and who wished us distinctly to understand that he did not claim to produce the phenomena he hoped to show us, by spirit agency. Prior to placing the ordinary slate under the table, we had washed and examined a small double folding slate, also belonging to the medium. This slate was locked by Mrs. Sidgwick, who put the key in her purse, and the purse in her pocket, and who sat upon the slate.

The single marked slate, of which I have previously spoken, was now held by Mr. D. under the table. We joined hands, and contented ourselves by asking merely that any word might be written, or the single word "Abbey." The medium, after the lapse of a few minutes only, showed himself very impatient at no writing having been produced. He proposed using another slate, which had also been washed and dried by us. Another piece of pencil was tried; these and other movements—for he constantly moved the slates to ascertain whether anything had been written—made it much more difficult to watch him narrowly, than had he been content to wait quietly and patiently for results. Still I was notable to detect any change of slates beyond the two which had been washed and cleaned by us. No writing had appeared, and Mr. D. soon proposed that we should try the following test, suggested by him.

[a] One of us was to stand with a newspaper on a table behind us, and with one finger was to point at random at any word; the other was to sit at the deal table with him. It was agreed that I should be the one to point to a word on the newspaper behind me, and I took up my position for this purpose at another table, about a couple of yards from the one at which Mrs. Sidgwick and Mr. D. were sitting. We had waited a few seconds only when the medium—who, as I have said before, seemed restless and impatient throughout the evening—suggested that I should blow out the candles behind me, as he thought darkness behind the paper might facilitate the accomplishment of the test. I complied, and returning to the table again placed my hand behind my back, and put my finger, as before, at random on any part of the paper. I had no sooner done as he had asked, than Mr. D. regretted that I should ever have moved from my place, as he thought it possible that the word written—if writing came at all—would be the first and not the second word to which I had pointed. It was not long before writing was heard. Mrs. Sidgwick and Mr. D. turned to the candle behind them, but were unable to decipher the word written. Mr. D. asked that it might be re-written. Another slate was used, on which they presently read the word "Melbourne." We then turned to see the word at which I was pointing. It was, however, not "Melbourne," neither was that word anywhere in the immediate neighbourhood of my finger, though we did afterwards find it in larger type, and two columns further on. Whether I had at first pointed to this word, could not of course be ascertained. It is possible that I did so, though a test which we tried later on, and to which I shall presently allude, proves that the "control" was not incapable of making a mistake. As to whether this word was obtained under good conditions I cannot give an opinion, as I was not sufficiently near the medium to be able to watch him closely.

[b] We next reversed our positions, Mrs. Sidgwick sitting opposite Mr. D. and I next him; the double slate remained on the chair on which I sat.[1] After again washing two slates, and placing one on the

top of the other, Mr. D. and I together held them under the table; they were once or twice removed to see that the pencil was still there, and that no writing had come, but always returned immediately, and they were not reversed or changed by Mr. D., so far as I know, whilst we were together holding them, up to the time when we again apparently heard the scratching of the pencil. We found, on carrying the slates to the light, that there was a message of moderate length covering half the slate, and signed "J.S."

[c] Again we sat as before, and under the same conditions writing came on one of the two slates held by Mr. D. and me under the table, half the message being in red chalk and half in blue—a bit of each had been placed on the slate. Again I could detect no trickery whatever; the slates were clean when we laid the bits of chalk between them, and one was covered with writing when they were removed from the table. Unfortunately the slates had not been marked, and I did not notice whether the bit of chalk was resting at the last stroke of the last word, or whether the nib was at all worn down. The message was of decided interest, it was as follows:—"My dear Friends, We have no wish to deceive you as regards the Agency question. The name of a deceased relative, especially a mother, is [far too sacred, although] most effective," &c.—the ordinary Spiritualistic jargon—then, "Our friend made a mistake the other night, dear Miss S. Agradezco á V. su visita. Espero que le volviere á ver á V. pronto. Tengo que macharme, A Dios." Then the message continued: "Yes, your haunted house was a failure." The last part of this sentence was too badly written to be deciphered; it appeared to be to the effect that the writer had not been able to be present. Now this message appears to me to be *striking*, though not of course conclusive. I must explain that about a month ago I had had a sitting with Mr. Eglinton for slate-writing, and had asked a question in Spanish, to which I had had the reply that "there was no French scholar present." Hence the reason of the remark "Our friend made a mistake the other night," &c. (the former séance had taken place in the day-time). This message, which had amused me at the time, had by no means been forgotten by me, although it had not been in my conscious thoughts for many days, and certainly not during the séance of which I write; it was known to Mrs. Sidgwick also, to whom I had written an account of the séance with Mr. Eglinton soon after

it took place. But what caused an allusion to this séance of a month ago, and through the agency of a medium who it seems improbable could have heard of this former séance? However, as 1 have said before, interesting and striking though this message appears to be, it gives no proof of thought-transference, as Mr. D. may have heard of the séance through some ordinary means, though from his conversation he appeared to be on any but intimate terms with Mr. Eglinton, and he was quite unknown to me and I believe to Mrs. Sidgwick (?) before we met him on this evening at Dean's-yard. He also told us that he did not know a word of Spanish. Still, it would have been no very difficult task to have got a few sentences written for this particular séance, and there was nothing in the message from beginning to end which might not have been written previous to the séance. But admitting all this, I was not conscious of any movement by which he could have changed a cleaned slate, which I was holding with him, for one on which the message was written. The allusion to the haunted house is less striking, for Mr. D. is known to Mr. Podmore, and might have heard of it through him, or many other sources.

[d] We next tried for writing on the locked slate. I must remark here that though we had sat on this slate during the greater part of the séance, we had not done so throughout. We had left it on the chair when we turned to the candles behind us to read the message. Mr. D. had quickly picked it up, and asked us not to lose sight of it, as he wished to preclude all possibility of fraud. He might, of course, in this moment[2] have changed the slate for one on which a message was already written, but the nature of the test we obtained, I think, negatives this supposition; besides which, before it was held under the table, Mrs. Sidgwick gave me the key, we unlocked the slate, found no writing there, and after the slate was again locked, I put the key in my pocket.[3]

It was now proposed by Mr. D. that we should try to obtain a line from a page of any book to be taken at random from Mr. Podmore's shelves. This was done by Mrs. Sidgwick, who took care only—at Mr. D.'s request—to select a book with good type. This book was shown to Mr. D., who opened it, looked at the type, and considered it sufficiently clear. Mrs. Sidgwick placed it on the table, and her and my hands rested on it, whilst Mr. D. and I held the small locked slate under the table. It was at this point, after choosing her book, that the slate had been opened, found clean, and the key given to me. It was decided that Mrs. Sidgwick should think of the page of the book from which the line was to betaken, and I of the line, counting from the top of the page, it being agreed—at Mr. D.'s wish—that to facilitate the test, we should each think of a number below 10.

Again, so far as I could see, we gave Mr. D. no opportunity for changing the slate. I am quite certain that he did not do so whilst we were holding it together. And in this case the message must have been written in our presence, as we did get a line copied from this very book, though not the line of which we were thinking. When the slate was again unlocked, we found writing on each side; the message was to the effect that we were not sufficiently *en rapport* with one another to get the best results as yet, but that they were willing to give us some proof of their power. Then followed a few words in inverted commas, after which an illegible word, with which the message broke off abruptly. Mrs. Sidgwick then explained that she had been thinking of page 9, and I had thought of line 4. Mrs. Sidgwick quickly turned to this page and line, but no such words as those quoted were to be found. Mr. D. suggested that the 9 in Mrs. Sidgwick's mind might have been reversed and wrongly read as 6. We, therefore, turned to page 6, and on the last line of that page and the first line of page 7, we found the words for which we were looking.

In this case—admitting the genuineness of the phenomenon—there might again have been thought-transference, for the book had been in both Mrs. Sidgwick's and Mr. D.'s hand, and either might have caught sight of these very words. With this the séance ended, as Mr. D. expressed himself too tired to sit any longer, and complained of a very bad headache. He seemed to suffer much after each message had been produced, and complained of great dryness of the throat.

Jessie H. Symons.

November 19th, 1885.

SITTING XVI.

Report of MISS SYMONS.

Slate-writing séance given by Mr. D., at 14, Dean's-yard, Westminster, Mrs. Sidgwick and myself present:—

I took slates with me—two ordinary ones, and one a folding slate, framed in wood, with a padlock and key.

[a] We first used the ordinary slates; they were cleaned, dried, and placed one on the other upon the table, a nib of pencil between them, and Mrs. Sidgwick's, medium's, and my hands resting on them. No writing being heard, Mr. D. and I held them underneath the table. Eventually, however, writing was produced whilst the slates were on the table in position I have before described. The message was a long one, covering completely one side of slate. We examined them when they were placed the second time on the table, and satisfied ourselves that they were clean. I am sure that the slates were not changed, because mine had rounded corners and Mr. D.'s, I observed, were square.[1]

[b] The medium next asked me to fetch a book from the outer room. I took one at random from the shelves of the library. Mr. D. saw me take it out, but did not touch it. I brought it into the inner room and put it on a chair between Mrs. Sidgwick and myself, whilst we prepared another slate and bit of pencil. Being again satisfied that the slates were clean, the book—into which I had not looked, and the name of which I did not know—was placed on the slate, all our hands resting on it as before. I mentally thought of a page and line, from which a quotation was to be made, both numbers, at medium's request, being under 10. After a short time writing was heard. On the slate was written, "Cantor lecture will be given on Mondays at the Kensington Museum—this is all we have power to do." We looked at p. 2, line 7, the numbers I had thought of, but did not find the words quoted. The medium, however, was very sure that they would be found somewhere near, and he soon discovered on last line of p. 7 "Cantor lecture," and on second line of p. 8 "will be given on Mondays," and a few lines further down, "at the Kensington Museum."

[c] The test having been only a partial success, the medium proposed that we should try it again. He asked me to fetch a second book

Experimental Investigation

from the outer room. I took up a *Journal* of the Society lying on the table. I did not look to see which number I had chosen. Medium asked me to think again of two numbers under 10, to determine page and line from which quotation should be made. I did so, and very shortly after was written in red chalk, "No such page." This was true, for on opening volume we found it commenced at a hundred and something.

[*d*] Mr. D. wished to try this test again, so I fetched a third book. This happened to be *Time*—both he and I saw the title. This time I told him which numbers I was thinking of—p. 8, line 5. We held one slate under the table, and another with the book on it remained on the table—both these slates were Mr. D.'s. After a time writing was heard, and it was on the upper slate that we found the quotation, correctly given this time, "The Imperial Parliament," line 1, and then a few words taken from line 5. The slate used was a large folding one, with a lock, belonging to medium. Into this he slipped a sheet of paper and a bit of lead pencil; it was on the paper that the quotation was written. Mrs. Sidgwick had the key, and it was she who opened the slate. The séance was held by full gaslight. The writing came always on underneath surface of slate—that lying nearest the table.

[*e*] Mr. D. then proposed showing us another trick.[(3)] He took up 12 squares of paper, asked me to name any 12 animals I liked, whose names he wrote on the 12 squares of paper. These were shuffled together, and I was asked to choose one, which I was to glance at and then instantly to burn. Mr. D. at the same time threw the other squares into the fire. I next wrote the first and last letters of the animal I had chosen on another piece of paper, this Mr. D. burned in the gas, bared his arm and showed us that there was nothing written there, rubbed the ashes of the burnt paper over the bare arm, and presently what looked like letters became very faintly visible. They did not, however, become sufficiently distinct to enable us to read them, and Mr. D. said he would presently get the animal's name written on a slate.

[*f*] We sat round the table again, as before—Mrs. Sidgwick opposite medium and I next him. One slate was held underneath table by Mr. D. and me, and the others were left on the table, with our hands resting on it. I asked that the names of my sisters might be written—this was not done. Neither did "Joey," in answer to Mrs. Sidgwick's question, succeed in telling us where we had spent the greater part of the day. "Joey" was

also unable to get any writing on my little folding locked slate, though we gave him two or three times the opportunity of doing so.

[*g*] Mr. D. asked him to tell us any secrets about either of us, and we heard the sound of writing on the slate lying on the upper surface of the table. The sound continued, when Mr. D. withdrew his hands a short distance from the slate, but ceased when he withdrew them to a greater distance. A long message was written again, covering the whole side of the slate, and commencing at a spot where the medium had previously requested it to commence by putting a small cross.

The "secrets" were such as were more or less known to us all, referring to a possible explanation I had given of our last séance with Mr. D. to "Blue Bricks," and telling me that I could get slate-writing if I sat sufficiently often, and "not with Mrs. Sidgwick"!!

One of our messages at request was written in different-coloured chalks, three bits of which had been placed on the table, underneath the slate.

[*h*] The last experiment Mr. D. showed us was the visible moving of the chalk under an inverted tumbler.

Two bits of chalk were placed on a slate, the tumbler covering them; the slate was isolated from the table. Mr. D. held Mrs. Sidgwick's and my hands, and there was no contact by either of us with the slate. Presently one bit of chalk was observed to move slightly. Mrs. Sidgwick asked it to trace the figure 4 (Mr. D. having proposed that she should choose a number under 10), and on removing the tumbler and inspecting the slate, we found the figure 4 somewhat faintly traced on the slate. I do not believe that this was a genuine phenomenon, though I have no theory as to how the trick was performed. I only observed that though the chalk moved, it did not appear to be forming a 4, although that figure was plainly visible when Mr. D. afterwards gave us the slate.

[*i*] Before he left, Mr. D. held a slate with me under the table, and asked that the name of the animal written on the slip of paper I had chosen should be written on the slate. Writing was heard, the slate brought up, and I found "rhinoceros" —wrongly spelt—in red chalk. This was correct, though how Mr. D. knew, or by what means the word was written, I have no idea, for the slate appeared to me to be clean[4] when we put it under the table.

JESSIE H. SYMONS.

February 23rd, 1886.
SITTING FOR MATERIALISATION.

The foregoing reports have all related to "Slate-writing," or analogous phenomena. I have, however, also given a few sittings for "Materialisation," and I may in the future endeavour to exhibit more fully the possibilities of trickery in this direction, but this branch of the subject has been of less interest to me, partly because the experiences which originally impressed me in connection with Spiritualism were not "Materialisation" but "Slate-writing" phenomena—partly because the testimony offered by Spiritualists for the genuineness of the latter appears to be so much superior to that offered in favour of "Materialisations." Still, the following reports of a séance which I gave last year may be instructive by way of suggesting what may be done by trickery. Although only three of the six sitters wrote reports, none of them contributed in the smallest degree to the production of the "phenomena."

1. *Report of* MRS JOHNSON.

October 7th, 1886.

I have just returned from paying a quite unexpected visit to Mr. Davey. We had invited him to our house for tomorrow to give us one of his wonderful manifestations, but received a bad account of his health, which prevented him keeping his appointment. I, therefore, with my sister, called to inquire after him, and found that, although unable to leave the house, he was about to hold a séance with some friends, and invited us to join them.

On entering the dining-room we searched every article of furniture, but could find nothing that could in any way assist in the materialisation which followed. Mr. Davey also turned out his pockets, and we looked under his coat and waistcoat. After the door was locked and sealed, and the gas turned out, we, six besides Mr. Davey, sat round a table, all joining hands. I had hold of Mr. Davey's left hand and a gentleman opposite of his right, none for a moment letting go until the end of the séance. A musical box was playing on the table; by degrees it floated about and knocked a gentleman on the head. Knockings were heard in different parts of the room, and bright lights seen. A gong sounded several times, and then appeared the head of a woman, which came close to us, and then dematerialised. After a few seconds another form appeared, the half figure of a man holding a book, with lambent edges, which it raised over its head, moved close to us, bowed several times, and by degrees seemed to disappear with a scraping noise through the ceiling. During the séance I with the others had various taps on the head and body, a gentleman complained of the coldness of a hand pressing on him, and the séance was altogether a most interesting, remarkable and startling phenomenon, and I can in no way account for it.

MARIANNE JOHNSON.

2. *Report of* MISS WILLSON.

Thursday Evening, October 7th.

DEAR MR. DAVEY,

We have just returned from a séance at your house, and while all is fresh in my memory, I hasten to send you my account of what happened. You had kindly promised to come to my sister's house tomorrow evening to give us a "materialisation," and perhaps some slate-writing, but having received a telegram and letter from you saying you must disappoint us, as your doctor had forbidden you for the present to exert yourself much or to be out in the night-air, my sister and I called today to inquire after you.

We found you at home, and you persuaded us to stay to a short séance.

Seven of us, including yourself, entered the dining-room, which we immediately examined, looking under the tables and sofa, behind curtains, inside the cheffonier, &c. After convincing ourselves that nothing was concealed, and you having turned out your pockets, we locked the door, and placed a sealed paper across it. We then sat round the dining-table, holding hands in a circle, a musical box was placed on the table, and the gas turned out. In a short time we heard raps in various parts of the room, a gong sounded in one corner, the musical-box, playing, floated in the air, and struck the head of one of our party. Several felt themselves touched, and one said he distinctly felt a cold hand placed on his head.

A female head appeared, in a strong light, floating in the air, and afterwards a half-length figure of a bearded man, in a turban, reading a book, appeared in the same manner, bowed to some of the assembly, raised his book above his head, and floated about the room, finally disappearing through the ceiling with a scraping noise. This all happened while two of our number tightly held your hands, and are convinced they never relaxed their clasp.

On the gas being relit, we found the door still locked with the paper unbroken.

Trusting our visit did not fatigue you, and that your proposed trip will soon restore you to health, when we hope you will resume your interesting investigations, I remain, sincerely yours,

E. M. WILLSON.

3. *Report of* MR. RAIT.

On Thursday evening, the 7th October, 1886, I was present at a séance held by Mr. Davey, at his house. There were in all eight persons, myself included.[38] We took our seats at 7.30 p.m., round an ordinary dining-room table (in the dining-room of the house), which, at Mr. Davey's request, we examined carefully, as also any other objects in the room which demanded our attention. The door of the room was locked, and I placed the key in my pocket, it was also sealed with a slip of gummed paper; the gas was then turned out, so that we were left in darkness. A musical box was wound up, and set to play an air, with the object, as I suppose, to enliven the proceedings! I held Mr. Davey's right hand, his left was held by Mrs. [Johnson]; the rest joined hands, so that during the séance a continual chain was formed which was maintained the whole time. After we had remained some time thus, various noises as of a shuffling of feet, &c., were heard in different parts of the room, and I distinctly felt something grasp my right foot; almost immediately I was touched on the forehead by a cold hand, which, at Mr. Davey's request, also touched those that wished it. The musical box was lifted, and although it was dark I fancied I saw it, surrounded by a pale light, descend through the air; it certainly struck me lightly on the side of the head, then it was again raised, and deposited on the table.

---

38 *Seven.* See the other reports.—S. J. D.

The Possibilities of Mal-Observation, &c.

The hand which touched me was cold and *clammy*; it evidently belonged to a most courteous and obliging spirit, for it did exactly what we desired! and at my wishing to feel the full palm on the back of my head (so as to ascertain its shape and size) it rested there for fully three seconds; it was, however, a somewhat weird experience! Various raps were now heard, a gong sounded behind my back, and we were told by Mr. Davey to pay attention, as something wonderful was about to take place. Faintly, but gradually growing more distinct, a bluish white light appeared hovering about our heads; it gradually developed more and more till at length we beheld what we were told was the head of a woman. This apparition was frightful in its ugliness, but so distinct that every one could see it. The features were distinct, the cheek bones prominent, the nose aquiline, a kind of hood covered the head, and the whole resembled the head of a mummy. After favouring those of the company who wished to see its full face by turning towards them, it gradually vanished in our presence. The next spirit form was more wonderful still; a thin streak of light appeared behind Mr. Davey, vanished, appeared again in another part of the room, and by degrees developed into the figure of a man. The extremities were hidden in a kind of mist, but the arms, shoulders and head were visible. The figure was that of an Oriental, a thick black beard covered his face, his head was surrounded by a turban; in his hands he carried a book which he occasionally held above his head, glancing now and then from underneath it. The face came once so near to me that it appeared to be only two feet from mine. I thus could examine it closely. The eyes were stony and fixed and never moved once. The complexion was not dusky, but very white; the expression was vacant and listless. After remaining in the room for a few seconds, or rather a minute, the apparition gradually rose, and appeared to pass clean through the ceiling, brushing it audibly as it passed through. The séance here terminated; the gas was turned on again, and everything appeared the same as when we first sat down; the door was unlocked, the seal being found intact. I will mention that during the whole of the séance I held Mr. Davey's right hand, with but one exception, when it was found necessary for him to light the gas to see to windup the musical box, as it had stopped playing. Nothing was prepared beforehand; the séance was quite casual; we could have sat in any room we wished, and we had full liberty to examine everything in the room, even to the contents of Mr. Davey's pockets, which were emptied (before beginning the séance) by him on the table before our eyes!

JOHN H. RAIT.

October 8th, 1886.

Now I should have no hesitation whatever in challenging Spiritualistic "mediums," or any other persons, to reproduce the phenomena described in the various reports which I have quoted, *under the conditions described by the witnesses.* I need hardly say that not one of these detailed reports is accurate throughout, and that scarcely one of them is accurate in even all the points of importance. I think it undesirable at present to publish the details of my methods, but I have communicated them to Mrs. Sidgwick and Mr. Hodgson. I have also communicated them to Mr. Angelo J. Lewis, known under the name of Professor Hoffmann as the author of several books on conjuring and magic. Mr. Lewis sends me the following statement:—

I have read with much interest the foregoing reports of sittings with Mr. Davey, testifying, as will be seen, to occurrences fully as striking and apparently abnormal as anything recorded as having taken place at sittings with Mr. Eglinton. I have since had the opportunity of discussing the matter in detail with Mr. Davey, who has indicated how far the descriptions of the sitters (though given in all good faith) differ from the actual occurrences, and has explained the various methods employed by him, some of such methods being those in actual use by professional mediums in America and elsewhere, and others the outcome of his own ingenuity. I have been much struck with their combined boldness and simplicity, and in view of the complete illusion they admittedly have produced in so many cases, the "doubt" which I expressed in the *Journal* of the Society for Psychical Research for *August,* 1886, as to the possibility of the whole of the Eglinton manifestations being produced by trickery, has been greatly shaken. Mr. Davey's successes prove that it is possible for a conjurer, devoting himself specially to slate-writing feats, to produce, under the same external conditions, results of precisely the same kind and quality as those produced by the professed medium. Indeed, in so far as the conditions vary at all, they are greatly in favour of the professional medium, first, by reason of the prestige derived from his claim to supernatural powers; and, secondly, by reason of his cherished privilege of producing no results at all unless he may consider it perfectly safe to do so.

I am not at liberty to divulge Mr. Davey's methods, nor would any good purpose be served by doing so; but I willingly certify, for the benefit of any person who may still entertain a doubt upon the matter, that his "manifestations" are in every case produced by perfectly natural means, no Spiritualistic or other unknown force having any part in them.

ANGELO J. LEWIS.
("Professor Hoffmann.")

Enough has been said, in the notes to the reports, to suggest to the reader how wide a margin must be allowed for the possibilities of misdescription in the numerous records to be found in Spiritualistic literature, of occurrences described in much the same manner as those which I produced by trickery. It seems to me not improbable that had I claimed the agency of "spirits," the effect upon many of my sitters would have been yet more impressive, and their reports would have been still more wonderful. But my position debarred me from more than one advantage which has been used, I believe, by many a trickster "medium." I was unwilling, for instance, to trade upon their emotion by professing to give messages from dead relatives and friends, as I might in many instances have done thus rendering the recipients of such messages less capable as observers of the phenomena, and more prejudiced in favour of their genuineness. I should find it very difficult myself to draw any line as to the possibilities of mal-observation and lapse of memory in *bona fide* witnesses beyond saying that I should allow for *at least* as much as may be exemplified in the foregoing reports.

I may here again remind my readers that it was not as a sceptic, but as a believer in "psychography," that I originally approached this investigation as to the results that might be produced by conjuring. I gradually became convinced that my belief in "psychography" was unjustified, that I could not attribute any value to the records of sittings which I had with a professional medium in 1884, without claiming a superiority which undoubtedly does not exist, to the numerous witnesses of my own phenomena.

In some of my earlier experiments I believed that there were indications of thought-transference between myself and my sitters. My later sittings have offered no support to this view, but, owing partly to my inexperience, I laboured sometimes under considerable nervous excitement in my earlier sittings, and I have not felt this latterly. This may have conduced to what occasionally seemed to me to be a certain amount of community of thought between my sitters and myself, and I hope at some time to make a special series of experiments for the purpose of ascertaining whether my conjecture is well founded or not.

In conclusion, I may say that the results of my investigation as to the possibilities of conjuring in relation to "psychography" have been a revelation to myself no less than to others. I am aware that in addition to the methods which I have employed for producing "slate-writing," there are other methods, which I know to be conjuring, but which have not yet been shown to me; and I should certainly not be convinced of the genuineness of Spiritualistic phenomena of this kind by any testimony such as I have seen recently published in great abundance, which presents so many close analogies to the reports of my own conjuring performances.

APPENDIX.

Notes to SITTING I.

By S. J. Davey.

[From notes made September 19th, 1886.]

1. Although Mr. Rait's slates did not leave the room during the séance, one of them was left unguarded on the table on one occasion for about sixty seconds.

2. The chalks *had* been taken out of the box before the séance.

3. This was not invariably the case; Mr. Rait examined the chalks on only two or three occasions.

4. *I* put the slate below the table, and after a while I asked Mr. Rait to help me to hold it.

5. *Immediately* should be *about four minutes after the question was asked*; another writing was produced in the interval. See Mr. Limmer's report [c.c.].

6. It was *I*, not Mr. Rait, who suggested the change.

7. There is no mention, in either account, of my manipulations of the slates after the experiment [*g*] was decided upon.

8. Mr. Rait omits to mention that this question had been asked early in the sitting. See Mr. Limmer's report [*e*].

9. At the conclusion of the message Mr. Rait opened his envelope and I saw "September," and was therefore able to impart this information on a slate later on.

10. I "led up to" this request by Mr. Rait.

11. It is obvious that a few words or sentences, or, if required, a long message, can be produced on special occasions, from languages unfamiliar to the "medium."

12. Mr. Rait did not take proper precautions for identifying the pencil.

13. It was a single slate that was used in experiment [*l*]; see Mr. Rait's report.

Notes to SITTING II.

By Richard Hodgson.

[From notes made September 17th, 1886.]

1. Mrs. Y. does not mention that the slate was withdrawn more than once and placed on the table, and she apparently did not observe what D. did with the slate on these occasions in the act of placing it under the table again, and before Miss Y., who relinquished hold when the slate was on the table, again took hold of it.

Mrs. Y also does not say what the questions were. One of them, asked by Mr. Y., was, "On what day do we sail for America?" and another, asked by herself, was, "Have I got over-shoes on?" It is noteworthy that *later* a reply *was* obtained to this first question, which D. requested Mr. Y. to repeat.

2. It was, however, written in an ordinary way. It was not the case that neither Mr. D.'s hand nor Miss Y.'s moved in the least the whole time. Part of the conjuring operation took place while Miss Y. was holding the slate, and while the thumbs of D. and Miss Y. were both visible, but another part of the operation took place in the intervals when Miss Y. was not holding the slate.

3. Mrs. Y. might have added that three more candles were burning on the mantel-piece, and a lamp turned to the full on an adjoining table.

4. I incline to think that D. walked with Miss Y. close to the bookcase, and waved his hand, requesting Miss Y. to choose a book.

[I did not remember this incident clearly when I made my note, my attention having been drawn elsewhere while Miss Y. was making her selection. But I learn from Mr. Davey that I saw correctly. I remember that after the writing had been produced, Mr. Y. asked Miss

The Possibilities of Mal-Observation, &c.

Y. if she had gone alone to the bookcase, and she replied that she had, and that Mr. Davey had remained by the table with his back towards her. Hence, probably, the agreement of the reports in the erroneous statement. I conjecture that Miss Y. transposed Mr. Davey's actions on the two separate occasions of her choice of a book. On the occasion of her first choice I believe Mr. Davey did remain close to the table as she describes.]

5. Miss Y. had thought of page 1, but no quotation had been given from this page.

6. This statement is erroneous.

Mrs. Y. had not the slate under her eye the whole time, nor was it the case that either her daughter's hand or her own was placed upon it continuously.

7. This statement also is erroneous. The slates used in the experiment were, indeed, those which Mr. Y. had brought with him, but they were taken and placed together by Mr. Davey.

8. Mrs. Y.'s hand was removed from the slates during the experiment. 9. After the figure had been drawn and the name of the colour written in the locked slate (away from the table, and out of the sight of D.) by Mr. and Miss Y., Mr. Y. named the *colour* (only) aloud, at Mr. D.'s request.

10. Miss Y. does not mention the previous withdrawals of the slate. See note 2, and Mr. Y's. report [*a*].

11. Miss Y. herself took the slate in her own hand with her to the bookcase, and brought it back to the table, without relinquishing hold. She has thus here forgotten a precaution which she really took.

12. Miss Y.'s remembrance is here incorrect.

13. Miss Y. wrote her description of this from her recollection of the figure seen on the slate the night before, and she stated, after looking at the slate, that she had in memory confused the white and the so-called green (which in daylight is at once seen to be *blue*). The *long* mark was blue, the *short* one was white.

14. I think that two out of the three books originally chosen had been *previously* replaced, the one chosen by Miss Y. being one of them.

15. I chose mine by requesting Mr. Y. to take a number of chips of pencil at random out of the box, which he did, the numbers giving 7 and 9.

16. Each wrote down on a slate the page and line he or she had chosen, so that D. could not see it, and then placed the slate, writing downwards, on the table under his or her own charge. The numbers chosen were not spoken aloud till after the final opening of the slate.

17. The slate was not guarded continuously during the interval between the examination and the final opening.

18. I consider that the figure was a manifest attempt at a cross, three of the limbs being clear, the fourth being only a scrawl. The marks forming the "cross" were in blue and white; there was also a red track on the slate.

19. I have heard Mr. Y.'s explanations, but with the very partial exception of a portion of one of them they were incorrect, and I believe that there was no thought-transference.

The reader should remember that the object of the notes to the various reports is not, of course, to supply all their deficiencies, or even to point out all the errors and omissions noticed therein by Mr. Davey and myself. Some of these errors and omissions the reader may discover for himself by a comparison of independent reports of the same sitting. Let him compare, for example, the three descriptions of [*e*] in Sitting II., bearing in mind that Mr. Y. knew all along that the performances were conjuring, that Miss Y. knew this only just before writing her report, and that Mrs. Y. did not know it until her report had been written. I agreed with Mrs. Y. that some of the other chalks moved, besides the red piece, and that the figure produced on the slate was clearly intended for across (and Mr. Davey afterwards assured me that he had intended to draw across), and I have no doubt that Mr. Y.'s description of the incident unduly diminishes the "marvel" of the phenomenon, just as Mrs. Y.'s unduly increases it.

Notes to SITTING III.

By Richard Hodgson.

[From notes made September 13-15th, 1886.]

1. This was not the case. The slate was sometimes lying on the table, with Mr. Legge's hand upon it; and he lost perception of it for a short period, notwithstanding his vigilance.

2. Mr. Davey took the slate and asked Mr. Legge to hold it under the table with him. After a short time the slate was withdrawn by both of them at Mr. Davey's suggestion, but no writing was found upon it. When Mr. Davey, alone, took hold of the slate again to place it under the table, he used an opportunity, and Mr. Legge did not observe what was done. These circumstances are omitted from Mr. Legge's report. The withdrawal which in his report appears to have been the first, was in reality the second.

3. Mr. Legge did not act precisely as before. On the second and third occasions on which the slate was placed under the table, Mr Legge did not take hold of it until after it had been placed there by Mr. Davey.

4. Mr. Legge omits to notice that Mr. Davey had asked a question as to whether any manifestations could be obtained.

5. The slate was neither selected nor placed by Mr. Legge. Mr. Davey first placed some coloured nibs of chalk on the table just in front of Mr. Legge. He then took one of his own slates which Mr. Legge had not touched, and apparently sponged both sides thoroughly. Mr. Davey himself then placed the slate over the pieces of chalk, and asked Mr. Legge to place his hand upon the slate. Mr. Legge then for the first time touched the slate.

6. Mr. Legge has omitted to mention more than one important previous detail concerning the locked slate. After locking it, he first, at Mr. Davey's request, put it in his pocket, also the key. Later on, he was requested by Mr. Davey to bring it out and place it on the table and put his hand upon it, first opening it to see if any writing was in it, and locking it again, and taking possession of the key. It lay on the table thus some time, and Mr. Davey found an opportunity of manipulating it. (See Note 1.) Mr. Legge's great care over some precautions

was the very cause of his neglect of others. 7. In addition to the lamp, there were four candles burning the whole time. Three of them were on the mantel-piece. I do not recollect whether the fourth was on the mantel-piece or on one of the tables.

Notes to SITTING IV.

By Richard Hodgson.

[From notes made September 22nd, 1886.]

1. The slates were both of them Mr. Padshah's, but I cannot recollect that either of them was washed by any person, and I find upon inspection—as they are still in my possession—that they were certainly not washed.
2. Mr. Padshah has omitted to mention myself. I sat between Mr. Russell and Hughes.
3. I do not myself remember. Mr. Russell states that it was washed by Mr. Davey, and Mrs. Russell states that it was washed by Mr. Padshah himself.
4. I thought the selection at that time had reference to the writing on the slate held underneath the table.
5. See note 15.
6. Mr. Padshah does not say that it was Mr. Davey who suggested that writing between the two slates should be asked for.
7. The word really written, as I learn from Mr. Davey, was "Books," and was thus curiously misread by Mr. Padshah, who had, he tells us near the beginning of his report, suggested the importance of getting "my own name—not surname—which no one except myself in the room knew." I believe that no one in the room except Mr. Padshah knew that his initial name was *Boorzu*, in the original Persian. Mr. Davey had written "Books" in order to suggest the experiment with a book, which was afterwards carried out. Mr. Padshah had apparently been much impressed with the idea of getting his first name written, and it is no matter for surprise that, with such a dominant idea, he should in-

terpret a scrawly *Books* into a *Boorz*. It was just Mr. Padshah's devotion to his test that produced the illusion.

8. Not cracks, but a peculiar chip in the frame on one side, which I had observed closely when Mr. Padshah first showed me his slates.

9. Mr. Padshah had not examined them.

10. What Mr. Padshah describes as a "push" was merely the result of the shaking of Mr. Davey's hands in his endeavour to produce the appearance of "convulsive movements."

11. None of Mr. Padshah's colleagues expressed any opinion as to the manner of the production of the writing, and it was explained before the séance that the reports written should be as independent as possible.

12. Notwithstanding Mr. Padshah's confidence on this point, this "contemptible" theory is the true one; his attention was actually diverted from the locked slate, and for some time he entirely lost perception of it although it was then lying on the table. His confidence on this point is a striking illustration of the influence of temporary forgetfulness, which not improbably would have become permanent had he not, after giving me his report, made further efforts of recollection after I had told him that the slate-writing was due to conjuring. He wrote his report on the night of the sitting, beginning shortly after the sitting was over. He gave me his report as soon as he had finished it, and I then assured him that the slate-writing in every case was the result of conjuring, that the writing on the slate was not "precipitated," but was ordinary writing with a slate pencil, and that he had actually lost sight of the slate during the sitting. He then endeavoured again to recall the events of the sitting, and succeeded eventually in recollecting the particular occasion when an opportunity was given to Mr. Davey of dealing with the locked slate unobserved.

[I did not then inform Mr. Padshah whether he was right or not in this recollection, and his temporary forgetfulness has become permanent. Apparently he afterwards quite forgot the occasion when he gave Mr. Davey the opportunity to produce the message, and he wrote to me on November 21st, 1886: "I now imagine Davey wrote while he studiously directed my attention to a variety of books, and naturally absent-minded, my attention was absorbed with a view to make a judicious selection, thus withdrawing my eye from the slate itself. Whether it is so or not, I think it is more than possible it might have been done that way." During this incident, however,—unfortunately for Mr. Davey—Mr. Padshah had taken the locked slate with him and carefully guarded it; see Mr. Russell's report [*g*].]

13. I learn from Mr. Padshah that he has had sittings with Eglinton when "phenomena" occurred, that he was not convinced by them "of something abnormal," and that he was much more impressed by the sitting with Mr. Davey.

14. Mr. Russell took very little share in the talking, and can be hardly said to have joined in the conversation at all!

15. Mr. Russell here states that the "Yes" and the "6" were found at the same time on the slate. My remembrance as to this point is not vivid, but it agrees with Mr. Russell's. From Mr. Padshah's report it would appear that there were two withdrawals, the first for the "6," the second for the "Yes." Mrs. Russell also makes two withdrawals, but reverses the sequence, taking the first for the yes, and the second for the "6." The circumstances occurred, I think, in the following order:—

Mr. Padshah was in the first place asked to think of a number; later, Mr. Davey put the question as to whether there would be any manifestations. When the slate was withdrawn, the *yes* was first *observed*, then the 6.

16. Mr. Russell makes this statement as though he had read and distinguished the letters himself, which I believe was not the case. [I have since learnt that he did not see the word at all, as it was so hastily wiped away by Mr. Davey.]

17. I think that Mr. Davey turned the two slates over together in the act of placing them on Mr. Padshah's shoulder, not in the act of replacing them on the table. [Mr. Davey tells me that Mr. Russell is right; Mr. Hughes agreed with me.]

18. *Blue.* See Note 22.

19. Mr. Russell made a very few brief notes during the sitting, but did not use these in writing his report.

20. There is a drawer at *one end only* of the table. Mrs. Russell probably *inferred* that there was a drawer at the other end, where, however, the table has never been fitted for a drawer. The table is perfectly honest, and the drawer has never been used by Mr. Davey.

21. It was Mr. Russell who chose red, and Mrs. Russell adopted his choice.

22. There were chalks of five colours between the slates, red, green, blue, yellow, and white. The "blue" writing afterwards exhibited appeared in the then light to be of a greenish tinge, and as attention was drawn to the writing's being green, Mr. Davey abstracted the blue piece of chalk that had been between the slates, probably so that if investigation of the chalks were made, it might be said that as there was no blue between the slates, green had been used. In daylight the writing was at once seen to be blue. Mr. Davey's manipulation of the chalks was not observed by any of the other sitters, and I mention it as typical of many incidents which occurred at the sittings where I was present, and which in some cases were of the utmost importance, but were entirely unnoticed by even the keenest of the uninitiated witnesses.

Notes to SITTING VIII.

By R. Hodgson and J. M. Dodds.

1. I had some conversation with Mr. Dodds on the day after he finished his report, and notwithstanding his close observation during the sitting, and the great care which he had taken to record accurately the events which he had witnessed, he agreed with me concerning particular lapses of observation and memory, which have produced some errors in his report. That he did not discover Mr. Davey's *modus operandi* in producing the writing was due chiefly to mal-observation, but mal-observation of a kind that perhaps the keenest uninitiated witness would find it almost impossible to avoid. Mr. Dodds at onetime or another had lost perception of each slate upon which writing was produced. One of the instances of lapse of memory is worth specifying because it illustrates tendencies to which I have previously adverted, —the tendency to minimise the marvel of a phenomenon known to be due to conjuring, and, possibly, the tendency to represent a subsequent impression as having been experienced during the sitting.

Experimental Investigation

The word *Yes* found written upon the slate was the word desired by Mr. Dodds himself, his question on the double-slate having been given up for the time for the express purpose of obtaining some simple phenomenon which was not to be regarded as a test, but merely as a ' start. Mr. Davey had suggested that we should try to get some simple word written such as no or yes, that if writing once began, we should probably "get plenty of it," and test questions could be attempted later. Mr. Dodds assented to this. Hence neither Mr. Davey nor myself expected Mr. Dodds to be specially influenced by the production of the word, though he appeared to be much more impressed in the first instance than he afterwards, when writing his report, supposed himself to have been.—R. H.

On talking over the sitting with Mr. Hodgson, two days afterwards, I agreed with him that in my account of the production of the word "Yes," my memory played me false, and I unconsciously minimised the result; and that his account given above is the correct one.—J. M. D.

2. This happened accidentally, and it was Mr. Davey who drew attention to it.—J.M. D.

3. I chose this book because I happened to have been reading another of *Taine*'s books in my chambers on the morning of the day of my sitting.—J. M. D.

4. At this stage, had Mr. Davey been a professional medium, he would perhaps have expressed surprise at the prematurely discovered writing, and passed it off as an unexpected production of the "spirits," remarking that the sound of the writing was not always heard by the sitters, and that even the medium himself could not always tell when it was being produced. It would have been difficult for Mr. Dodds to have explained how, under the conditions as described by him, the writing, covering a side of one of his own slates, could by any possibility have been produced by Mr. Davey himself.—R. H.

Notes to SITTING XIV.

By Richard Hodgson.

[From notes made March 26th, 1887.]

1. According to my recollection, Mr. Davey used both hands in placing the slates under the table, and again in the course of replacing them upon the table, but his left hand did not remain below the table in any instance for a longer interval than—I should name as a maximum limit—ten seconds.

2. The word "Commandment" ends at the edge of the slate, but it begins very nearly at the centre of the slate, and is written almost parallel with the longer axis of the slate, traversing rather more than half the length of the slate surface. The scrawled words "Seventy-seven" are written diagonally, very nearly across the centre of the slate, and the number 77 is in a similar position on the other side of the centre. The writing intended for Hong Kong is in a position somewhat to the right of the centre. The slates are still in my possession.

Notes to SITTING XV.

By Mrs. Sidgwick.

1. According to my independent notes (made Nov. 17th, 1885), the locked slate was at this period examined and was blank.

2. According to my notes "this moment" was of sufficient duration to give plenty of time and opportunity to write the message.

3. I infer from my notes and recollection that no examination of the slate was made at this period, for I had in my mind at the time two possible explanations of the trick, and any such examination would have been incompatible with either. Moreover, Mr. Davey assures me that from the way in which the trick actually was done he knows that the slate cannot have been examined at this point. Miss Symons must have transposed the examination which I record as having taken place earlier (see Note 1), to this period.

Notes to SITTING XVI.

By Mrs. Sidgwick.

1. According to my notes (made Feb. 23rd, 1886), Mr. Davey had one round-cornered slate among his, though I noticed, as I thought, decided differences between it and Miss Symons'. This is worth mentioning, as showing a difference of opinion on a point which we both thought we observed particularly. Besides omissions, there are at least two important positive misdescriptions, which I am not at liberty to specify further, in Miss Symons' account of the first incident of the séance. I well remember the impression which this incident made on me at the time. 1 could not make it out at all. I believe I thought it more puzzling than any professed Spiritualistic phenomena I have seen, assuming these latter to be conjuring. There seemed less possibility of its having been done by conjuring. The hypothesis of the change of slates, which at first did not seem to be possible—and which was, in fact, as Mr. Davey assures me, erroneous—never seemed plausible; only I was unable to think of any other explanation at all. The reason it puzzled me so much was that I thought I knew pretty well the possibilities of slate-writing, and there seemed to be no loop-hole here for any of them. It may interest the reader to compare my own account of the incident:—

"Miss Symons' two slates were held together on the table and under the table by her and Mr. Davey. Then one of Mr. Davey's square-cornered slates was substituted for one of them; then again removed and the two round-cornered ones again held, on the ground that though it might be easier to get writing on Mr. Davey's slate, it would be more satisfactory to get it on Miss Symons'. We waited a considerable time. Mr. Davey asked me to draw the curtains between the two rooms. Then we again sat as before; the two slates on one another on the table and our hands on them. The sound of writing was heard, and, presently, on looking between the two slates, one of them was found to be written on all over one side. I cannot remember every detail of what occurred, but the impression produced on my mind most distinctly was that one of Miss Symons' slates had been written on all over one side [the impression was so far true], and that there had been no possible opportunity for Mr. Davey to have done this. The slate seemed to have been on the table with our hands on it from the moment we had seen it clean. I do not know what happened while I drew the curtain, but cannot conceive its having been done then. Mr. Davey and Miss Symons still sat at the table, and even if there had been opportunity there was no time."

2. According to my notes it was for the first of the three book experiments that the paper and lead pencil were used, and Mr. Davey agreed with me. The paper was S.P.R. paper, and Mr. Davey tore off a corner for further identification.

3. "Another trick" was Mr. Davey's own expression.

4. The word was, however, then already on the slate.

MR. DAVEY'S IMITATIONS BY CONJURING OF PHENOMENA SOMETIMES ATTRIBUTED TO SPIRIT AGENCY.

BY RICHARD HODGSON, LL.D.

In the Introduction which I wrote (*Proceedings* S.P.R., Vol. IV., pp. 381-404) for the late Mr. Davey's "Experimental Investigation," conducted for the purpose of ascertaining "the possibilities of mal-observation and lapse of memory from a practical point of view," I pointed out that "to explain the tricks would in itself be of little advantage to the investigator of the 'physical phenomena' of mediums"; that other methods than those employed by Mr. Davey may be (and unquestionably are) practised; and in any case that explanations of the methods in use would hardly be likely to convince persons who have testified from personal experience to the genuineness of the "psychography" of well-known "slate-writing" mediums that such methods were used for the production of the phenomena which they witnessed. "They will scarcely," I said, "be likely to remember the occurrence of events which they perhaps never observed at all, or observed only partially and erroneously; which, whether correctly or incorrectly observed, they have afterwards continually misdescribed or completely forgotten; and which, in many cases, would be distinctly excluded by the acceptance of their testimony as it stands." The notes appended to the detailed reports quoted in the article referred to would, we thought, sufficiently show to the reader the several kinds of mistakes made by intelligent witnesses in recording their impressions of performances like Mr. Davey's, and would enable the student—not necessarily to discover in every case the exact *modus operandi* of the tricks, for this appeared to us to be of trivial importance, but—to appreciate the unreliability of human testimony under circumstances common to such performances. It was, indeed, my own personal opinion that on the whole it was advisable that the methods of Mr. Davey should be described in detail, as far as possible, though in many cases it would be difficult to explain verbally exactly what occurred so that the reader could enter fully into the situation. Mr. Davey, however, was strongly opposed to the revelation, and for various reasons. His chief objections, I believe, were that other methods than the ones which he employed had probably been used by pseudo-mediums, that new methods would doubtless be invented, that the description of his methods would interfere greatly with his projected plan of giving numerous additional sittings and obtaining further reports (in connection with which he proposed to explain his methods fully), and that many of his sitters would be annoyed at finding precisely how they had been deceived. Mr. Davey's death has removed the only argument—I may now freely say—which had special cogency in my own case, viz., his purpose to give another series of sittings, all of which should be attended by a person thoroughly familiar with his methods, and cognisant beforehand (so far as such cognisance was possible) of the precise things which he intended to do; this person was to write an account both of what was intended and what he witnessed; Mr. Davey was to supplement this account by his own statements; and these accounts were to be compared with the reports of the sitter in each case. The object, of course, in this projected later series was to emphasise still more forcibly the unreliability of the testimony so widely accepted, among Spiritualists, as adequate to establish the genuineness of the manifestations in question.

It appears, however, that the accounts of Mr. Davey's sittings published in Vol. IV. of our *Proceedings* are in themselves more than enough to demonstrate the affirmed unreliability of such testimony, and to justify the position originally put forward by Mrs. Sidgwick that the possibilities of mal-observation and lapse of memory must be absolutely excluded before the testimony to "slate-writing" and similar performances can be taken into further serious consideration. This is clearly shown by the communication which formed the immediate cause of this article, viz., the letter of Mr. Alfred Russel Wallace printed in the *Journal* of the Society for Psychical Research for March, 1891, in which he stated that Mr. Davey's performances "are claimed to be *all* trick, and unless *all* can be so explained many of us will be confirmed in our belief that Mr. Davey was really a medium as well as a conjurer." At the close of my Introduction to the reports of Mr. Davey's sittings, I asked the "experienced Spiritualist" to "point out exactly where the difference lies between 'Mr. Davey's performances' and mediumistic phenomena." Mr. Wallace has accepted this challenge in the name of "many of us";—there is no more illustrious name than his upon the roll of adherents to a belief in Spiritualism; and his reply is substantially a confession that he cannot distinguish between Mr. Davey's performances and ordinary "mediumistic" phenomena. But, strangely enough, as it appears to Mrs. Sidgwick and myself, and others who were familiar with Mr. Davey's devices, Mr. Wallace's conclusion seems to be, not that the analogous phenomena which have been reported about "mediums" were due to trickery, but that Mr. Davey's performances were "mediumistic"! The issue has changed. We are no longer asked to prove that this or that medium is a "trickster";—we are asked to prove that Mr. Davey was not a medium! Could any better evidence be offered that Mr. Davey's performances and those of certain professional mediums belong to the same class?

Now, I am not at all sure how far my explanations of Mr. Davey's devices will make clear to Mr. Wallace and the many others who agree with him, that every apparently "phenomenal" occurrence at his sittings can be accounted for by ordinary means. It is impossible to reproduce all the details of the sittings, so that the reader may have a faithful picture of the seemingly insignificant incidents that made the writing upon a slate on or under the table, or the turning over of one or two slates, or the substitution of one slate for another, or the secreting and carrying out of the room (to deal with at leisure) of one of the sitter's own slates, appear to the instructed and watchful observer so transparently easy. I should have much greater confidence did I know that these doubters of Mr. Davey's dexterity were familiarising themselves with such books as Professor Hoffmann's *Modern Magic* and *More Magic*, Mr. John W. Truesdell's *Spiritualism, Bottom Facts*, and a recent book published in America by Farrington and Co. (St. Paul, Minn.), entitled *Revelations of a Spirit-Medium*. Above all, I recommend these doubters to experiment for themselves. It may be difficult for them to obtain the assistance of a person like Mr. Davey, but they can at least study from books on conjuring the details of many performances commonly exhibited on the public stage, and by accompanying their uninitiated friends to the entertainment, and listening to their accounts of the tricks afterwards, they will be, I venture to think, considerably helped towards a proper appreciation of the misdescriptions usually given of such performances, and will perhaps begin to see the absurdity of attributing "mediumship" to Maskelyne, or Lynn, or Davey. In this direction at least the account of Mr. Davey's methods may prove serviceable.

I shall begin by giving a brief statement of the chief methods used by Mr. Davey and then illustrate his actual practice by describing in detail some of the most important occurrences at sittings where I was present myself. I shall then state what occurred, according to Mr. Davey, at the sittings particularly noted by Mr. Wallace as remarkable, and finally give the explanation of other incident which without such special reference might still remain incomprehensible. Had I foreseen my departure for America, and my continued stay here, and therefore the impossibility of my conducting such a later series of experiments, as I have mentioned above, I should doubtless have reduced to written record at the time the details of the sittings which I myself witnessed, as well as Mr. Davey's statements concerning the other sittings. As it is, I must depend upon my recollections, assisted by the contemporary notes published, in connection with the reports, in Vol. IV. of the *Proceedings*. With the regular methods employed by Mr. Davey I was, of course, very familiar, as he frequently practised them in my presence, and consulted me about variations of them. Further, I talked them over in detail with Mrs. Sidgwick and Professor Hoffmann, and was present at five out of the sixteen sittings reported, and saw them used. I questioned Mr. Davey at the time about all the incidents at the sittings where I was not present, and was perfectly satisfied with his explanations. I may add that I have seen similar methods used by "mediums" in America, as will be seen later from my account of a visit to the notorious medium Slade.

Referring to the reports, it will be noticed that the manifestations most frequent at Mr. Davey's sittings were:—(1) Writing on the upper surface of a single slate held against the underside of the table; (2) Writing on the upper surface of the under slate when two slates were placed together above the table; (3) Writing in Mr. Davey's locked slate. I shall describe the normal method used in each case; I say "normal," because differences between the sitters as to their attention, &c., together with other incidental circumstances, produced, in almost every instance, certain slight variations from the prescribed steps.

(1) The slate having been cleaned and placed near the edge of the table on top, with a piece of pencil or chalk upon it, Mr. Davey takes a thimble-pencil from a hip-pocket, and slips it on the end of a finger, say the third, of the right hand. A thimble-pencil is a tailor's thimble with a small piece of pencil (or chalk) fastened to it. He then draws the slate over the edge of the table, with the thumb of his right hand on top of the slate, the finger with pencil being tucked into his palm, brings the first and second fingers up to the under surface of the slate, and slowly slides the slate under the table, requesting the sitter on his right to hold the slate with him, and to keep it pressed closely up to the under surface of the table. The sitter does so. The slate is out of sight, but the thumbs of the holders are visible. The sitter, in response to Mr. Davey's suggestion, asks a question. Mr. Davey writes the answer noiselessly with his thimble pencil on the under surface of the slate, without the knowledge of the sitter. After an interval of waiting, he proceeds to withdraw the slate, ostensibly to see if anything has been written. He places it on the table, and by that time the sitter has let go of the slate. Nothing is found written (on the upper surface of the slate, where the sitter knows that the writing is to appear if it comes at all, and where alone inspection is made). Mr. Davey lifts the piece of pencil off, rubs the upper surface again with a cloth, then seizes the slate with the fingers uppermost, and the thumb underneath, raises the slate from the table and places it once more under the table, turning the slate over as it is going under the table, and just before pressing it against the under surface of the table, drops upon it again a piece of pencil from the table. The answer to the question is now on the upper surface of the slate, pressed against the table. He then reminds the sitter to hold the slate also, and asks that the question be repeated. After

a short interval, the sound of writing is heard, caused by Mr. Davey writing (for it is possible to write either with or without noise), on the now under surface of the slate, the answer to a question not yet asked, and which Mr. Davey may ask himself after the next insertion of the slate under the table. The slate is then withdrawn as before, the answer on the upper surface is read, that surface is cleaned by Mr. Davey; the slate is again placed under the table and turned as before in the process. And so on.

(2) Writing on the interior surface of one of two slates held together above the table.

One slate has already been written upon, during or previous to the sitting, and this lies, writing downward, upon the table. Mr. Davey gives two other similar slates to the sitter to examine and clean, asks him to place pencil (or chalk) on one of them on the table, cover it with the other, and place his hands upon them. Mr. Davey also places his hands upon them. After an interval of waiting Mr. Davey suggests looking to see if there is writing. The sitter removes his hands, Mr. Davey takes off the top slate and places it with seeming carelessness on one side close to where the third slate is lying, and after removing the pencil, say, from the other slate, and perhaps rubbing it again with the duster, which afterwards is perhaps thrown on the slate just removed, and placing some pieces of chalk again on the slate, he takes the third slate (writing already on the under surface) and places it on top. The sitter and Mr. Davey place their hands on the slates as before. After another interval of waiting Mr. Davey proposes to hold the slates in the air, or resting against the sitter's shoulder. The sitter raises his hands from the slates. Mr. Davey takes the two slates together, the fingers of his right hand above, the thumb below, and in lifting them from the table turns them both over together. This movement is probably completed by the time the sitter also takes hold of the slates. After a short time, a sound as of writing is heard, and when this is finished, the sitter lifts the top slate, and finds the upper surface of the lower slate covered with writing. But what produces the sound as of writing? Sometimes the fingernail of Mr. Davey on the under surface of the bottom slate, some times a movement of his knee to which is attached a piece of common slate-pencil, the ends resting in two small loops of rubber sewn on to his trousers. He chafes this piece of pencil against another piece attached to a fragment of wood from which project two fine steel points, by means of which he easily secures it to the pendent rim or the leg of the table. This was Mr. Davey's variation, I believe, on the idea suggested by the wedge-shaped clamp illustrated by Mr. Truesdell. (*Spiritualism, Bottom Facts*, p. 199.)

(3) Writing in the locked slate.

Mr. Davey has two locked slates precisely alike, i.e., as precisely alike as skilled workmanship could make them. In some cases a communication was prepared beforehand, and when a reply was not demanded to some specific question, a single substitution was all that was required. When a question was asked in the locked slate, two substitutions were needed. Thus, a question is written by the sitter in locked slate A. Mr. Davey substitutes locked slate B for A, opens A and answers the question (usually taking it out of the room for the purpose), and later on re-substitutes it for B.

"Well, but," I hear some readers say, "I want to know exactly how and when he makes these substitutions, and besides, how and when does he cover the side of one of the sitter's own slates with writing?" It is just these questions that are difficult to answer satisfactorily without introducing the whole *mise en scène*, so to speak, of the sitting. I think, however, that a tolerably fair conception may be formed by considering several of the reports and describing, as far as I can now reproduce them, the immediately connected details. But before doing so I shall describe Mr. Davey's usual method of substituting one of his locked slates for the other. This might almost be called his favourite device.

The first step was to engage the attention of the sitter on some other object. This was usually done by starting an experiment with another single slate or pair of slates. While the sitter was occupied in cleaning a slate, or examining pieces of pencil or chalk, or inspecting the writing that so "mysteriously" appeared on the ordinary slate, Mr. Davey was manipulating his "duster," a cloth which he used for drying the slates. This, after perhaps drying a slate with it, he would throw, apparently carelessly, over the locked slate on the table, and so as to hide this slate completely. Then, under cover, occasionally, of the use of his handkerchief, he would slip the other locked slate from his coat pocket or from beneath his waistcoat, slide it softly upon the edge of the table, and, bending over the table somewhat, with possibly one arm resting far forward on the table, so as partly to obstruct the view of the moving slate, push the slate softly forward till it was near the first slate concealed by the "duster." He would then sometimes boldly remove the duster with the first slate inside, and, below the surface of the table, slip the slate beneath his waistcoat, afterwards replacing the duster on the table. Sometimes after the second slate had been placed upon the table, he allowed the first slate, covered by the duster, to remain on the table also for a considerable interval, owing to the possible danger of removing it without detection. On one occasion it remained there, I think, for more than a quarter of an hour, until at least the conclusion of the sitting, when he gathered up his various articles into his bag. While the sitter was wondering at the long communication in the *second* locked slate, the first locked slate, under the duster, was lying within his reach on the table before him.

Let us now consider the above explanations in detail with special reference to Sitting II. (pp. 57-70.) For convenience of reference I here reproduce portions of the accounts:—

Mrs. Y.'s Account.

A piece of chalk was placed on one of our slates, and the slate was held tightly up against the underside of the table leaf by one of Mr. Davey's hands and one of my daughter's. Their thumbs were on top of the table, and their hands spread underneath on the underside of the slate. I held Mr. Davey's other hand, and we all joined hands around the table. I watched the two hands holding the slate without a moment's intermission, and I am confident that neither Mr. Davey's hand nor my daughter's moved in the least during the whole time. Two or three questions were asked without any sign of response. Then Mr. Davey asked rather emphatically, looking hard at the corner of the table under which they were holding the slate, "Will you do anything for us?" After this question had been repeated three or four times, a scratching noise was heard, and on drawing out the slate a distinct "Yes" was found written on it, the chalk being found stationary at the point where the writing ceased. As my eyes were fixed uninterruptedly on both my daughter's hand and on Mr. Davey's also, and as I certainly had fast hold of his other hand all the time, I feel confident he did not write this word in any ordinary way. This same result was obtained two or three times.

Mr. R. Hodgson

Miss Y.'s Account.

Mr. Hodgson brought us a little pasteboard box, in which were a number of small pieces of chalk of different colours. I chose two of these and placed them on one of our slates. We had all previously written either our names or our initials on that side of the slate. Mr. Davey slipped the slate under the edge of the table, I holding on to it all the time, and we held it flat under the table with our thumbs above the table. I held the slate very firmly against the table, and I am sure I did not relax my hold once. After waiting some time and asking various questions, we heard, or seemed to hear, the chalk moving on the slate. We drew the slate out, and on it was written "Yes," which was an answer to our last question. We again put the slate under the table, and, in order to be sure that nothing had been written on it, I half slipped it out again and saw that it was perfectly clean. After some more waiting, my father asked when we were to sail for America. The chalk again squeaked, and on drawing the slate out we found "the 18th" written very indistinctly. This happened not to be the date, which was the 15th.

There is no mention of the previous withdrawals in either of these accounts. Mr. Y., however, did remember them, and recorded them in his report.

At the first and second examination nothing was on the slate, and it was washed afresh, and soon the word "Yes" was found scrawled on the upper side of the slate as an answer to some indifferent question.

Mr. Davey did not venture to write the word at once, and did not produce his writings continuously. The first part of a sitting was often a time of tedious waiting, so that the vigilance of the sitters might become relaxed, and so that they might be accustomed to regard the withdrawals of the slate as having no special meaning, if, indeed, it should ever occur to a sitter that they were suspicious. After one of the early withdrawals, Mr. Davey, having previously written the word yes on the under surface of the slate, turned the slate over in the act of replacing it, and, of course, during this interval Miss Y.'s hand was not holding the slate. She had relinquished her hold when the slate was placed on the table. Mr. Davey then asked his own question, to which the yes was a proper reply. Similarly after one of the later withdrawals, Mr. Davey, having written "the 18th" on the under surface as a reply to a question previously asked by Mr. Y., and having turned the slate over in the act of replacing it, requested Mr. Y. to repeat his question. "On what day do we sail for America?" There upon the writing was apparently produced, and the answer exhibited proved to be relevant, though the date given was not correct. It would weary the reader were I to point out all the discrepancies between the reports which I quote, such as that Mrs. Y. speaks of "a piece of chalk," and Miss Y. speaks of "two" pieces as having been placed on the slate. It is important, however, to emphasise here that although the slate was several times withdrawn from under the table, and Miss Y. on these occasions relinquished her hold of it completely, yet there is not the slightest indication in the above accounts that Miss Y.'s hold was ever relaxed at all, or that there was a single withdrawal when nothing was found written upon the slate. These are instances of the complete omission, from the record, of circumstances without which the trick would have been impossible. They were due to lapse of memory rather than to mal-observation, since at the times of the withdrawals the sitters were doubtless aware of them. The turnings of the slate as Mr. Davey replaced it under the table were probably not observed; that is to say, it was not observed that his method of placing the slate under the table brought the unexamined surface to the top.

After this explanation I think that the reader will find no difficulty in seeing exactly how the similar "phenomena" recounted in the other reports, in connection with a single slate held under the table, were produced. He must supply, of course, the "withdrawals" and the accompanying circumstances, since these are completely omitted from nearly all the records, and where the withdrawals are mentioned there seems to have been no conception, in the mind of the witness, of their significance.

Proceeding to the cases of writing appearing between two slates above the table, I quote the three different accounts from Sitting II.

Mrs. Y.'s Account.

After a short rest, Mr. Davey asked us to wash two of our own slates and put them together, with pieces of chalk of different colours between, and all of us to reach across the table and hold them all together. This we did, and then Mr. Davey asked my husband to choose mentally three colours he wished used in writing. After all holding the slates closely pressed together for a few minutes, we placed them on the table, and Mr. Davey and I placed our hands on them while the rest joined hands. In a few moments the same sort of electric shock seemed to pass through Mr. Davey, and his hand and arm which were on the slates quivered nervously, and immediately a scratching noise was heard. He then asked me to lift one slate off the other, which I did, and found one side covered with writing in three colours, the very three my husband had mentally chosen. I am perfectly confident that my hand was not removed from the slates for one single instant, and that I never lost sight of them for a moment.

Miss Y.'s Account.

After this experiment, we put aside Mr. Davey's slate and took two of our own. We cleaned them, and placed on one a number of little pieces of coloured chalk. The second slate was put on the first one, and my mother and Mr. Davey held it above the table. Mr. Davey asked my father to think of three colours. We joined hands once more, and in a little while we heard writing between the slates. When we took one off, on the under one was written:—

In *red*, "We are very glad to be able to give you this."

In *white*, "We can do more yet."

In *green*, "Good-bye."

My father had thought of red, white, and *blue*. We could not be sure by the night light whether the "good-bye" was written in green or blue. But there was a piece of chalk on the slate that looked much more blue than the piece with which the "good-bye" was written.

Mr. Y.'s Account.

We next placed small pencils, in six colours, between two of my newly bought slates, marked by ourselves with our names written in pencil, without removing them from the top of the table, and the hands of some of the party were laid upon them for some minutes, after which they were held up in the hands of two persons. I had been asked to choose the colours in which the writing should be made. I mentally chose red, white, and blue, but did not tell my choice. After holding Mr. Davey's hand for some minutes, with my mind strongly fixed on these colours, the slates were opened, and we found, in the order I had mentally selected:—

Mr. Davey's Imitations by Conjuring, &c.

(Red) "We are glad to be able to give you this."
(White) "We can do more yet."
(Blue) "Good-bye."

The slates used were the three ordinary school slates which Mr. Y. had purchased on the way to the sitting, which was held at my rooms at Furnival's Inn. The experiment preceding this was with Mr. Davey's locked slate. While the sitters were still pondering over the writing that had appeared in the locked slate, Mr. Davey retired to an adjoining room, taking with him, under his waistcoat, one of Mr. Y.'s slates. He there wrote upon the slate in red, white, and blue, thinking that if Mr. Y. were asked to choose mentally three colours, he would be more likely to select these three than any others. (I believe that Mr. Davey usually had red, green, pink, blue, yellow, and white chalks at his sittings.) Returning to the room and the table he surreptitiously placed this slate on the table again, writing downward, pushed Mr. Y.'s remaining two slates, which we may call the *first* and *second*, forward, and requested that these should be cleaned thoroughly. After the cleaning, Mr. Davey placed some coloured pieces of chalk upon the first slate and covered it with the second. According to my remembrance, Mr. Davey then lifted the two slates a little from the table and asked all the sitters to join in holding them. After a short interval he suggested looking to see if there was any writing, and the slates were lowered to the table, the sitters removed their hands, and Mr. Davey took off the top slate (the *second*), showing the under surface of it where there was no writing, and placing it on the table close to the *third* slate. Moving the chalks slightly, to be assured that there was no writing, he "replaced"—not the second slate which he had just removed, but—the *third* slate, which already had the writing on the under surface. He then placed his hands upon the slates, and so also did one or more of the sitters. After another short interval, Mr. Davey suggested holding them up in the air; the sitters lifted their hands, Mr. Davey seized the slates, raised them, turned them over together and requested, I believe, Mrs. Y. to join in holding them. At this stage I think that the sitters all stood up and that Mr. Davey then called upon Mr. Y. to think of three colours to be used in the writing. Very shortly the sound as of writing was heard. When the sound ceased, Mr. Davey let go of the slates, and the writing was found on the upper surface of the lower slate.

It is probable that my remembrance of the scene even where it is clear and distinct is wrong in some points, and on others even my remembrance is not clear. I cannot recall very clearly, for example, at what point Mr. Y. was asked to think of three colours. He may have been asked earlier to choose mentally three colours, and the request may have been repeated later. But these points are unimportant for my present purpose, which is to show the reader how the trick was done. I witnessed Mr. Davey abstract the slate; I witnessed him in the act of writing the message in the adjoining room; I witnessed him return the slate to the table, and afterwards substitute it for the other slate, and I witnessed him turn both slates over together as he raised them in the air. These were the important points for me to watch, as I knew beforehand.

Now for the omissions in the reports. In the first place, Mr. Y. and Miss Y. refer nowhere in their whole reports to the fact of Mr. Davey's leaving the room. Mrs. Y. refers to it, but supplements her reference by stating that "the slates were all the time in full view on the table with the rest of us who remained behind"! If she could but have seen Mr. Davey's hurry and excitement in the other room while he was preparing the message on one of her own slates!

In the second place, there is not the slightest indication, in any one of the three reports, that the slates were separated during the experiment after they had once been placed together, yet they were not only separated, but the separation was used for the purpose of substituting a third slate for one of the two cleaned for the experiment. (All three slates, I believe, had been marked at the beginning of the sitting, but no special markings were made for this particular experiment.)

We may now turn to the other cases where this particular trick was performed, and notice certain variations in the details. In Sitting I., the reader will easily be able now to supply the omission made by the two recorders; the top slate was removed and the third slate substituted in a manner similar to that described above. But in this case Mr. Davey wrote on the slate *before the sitting began*. Mr. R. took three slates to the sitting, and he states that there "could not possibly" have been "any tampering" with them, "as during the whole séance they never for one moment left the room." This last assertion is true, but Mr. R. gives 7.30 p.m. as the time of his going to Mr. Davey's house, and the sitting did not begin till 8.30 p.m. In the meantime Mr. Davey had taken one of Mr. R.'s slates and Mr. R.'s box of chalks also into another room, and written upon the slate, and rubbed away the corners of some of the fragments of chalk and pencil, and brought them back to the room. What occasion was there for the intending sitters to watch their slates then? The sitting had not begun, and besides, Mr. Davey had, let us say, given them some interesting curios or remarkable photographs to examine while he excused himself for a few moments. Later on, the sitting begins. Mr. Davey takes the parcel of slates: "Ah! these are your slates, Mr. R. Very glad you've brought your own slates. If anything comes, you see, it's so much more satisfactory. We'll try first if we can get any writing on one of your own single slates. Better clean it thoroughly," at the same time holding up the top slate for Mr. R. to take, and of course he takes it. The slate upon which Mr. Davey has written is the lower one of the remaining two, the writing being on the under surface; and these two slates remain on the table in full view while the experiments with the single slate are in progress. This important lower slate, however, becomes completely forgotten, temporarily at least, when the sitters are preparing the other two slates for the second form of experiment, at which time, had there been any need, Mr. Davey might have written upon it again. Hence Mr. Davey's note to this sitting that "although Mr. R.'s slates did not leave the room during the séance, one of them was left unguarded on the table on one occasion for about sixty seconds." As a matter of fact, however, as Mr. Davey assured me, he wrote upon this slate before the sitting began.

The next instance of this "manifestation" occurs in Sitting IV., held in my rooms; Mr. Padshah had taken three new slates, but had left them in my rooms while he made a call upon some friends. In the meanwhile Mr. Davey arrived, and used the opportunity to write on one of Mr. Padshah's slates, which he then placed at the bottom. When we were ready for the sitting, Mr. Padshah having returned, Mr. Davey began by passing round his locked slate for inspection, cleaning, &c. He then took the top one of Mr. Padshah's slates for trials with a single slate held under the table, leaving the two other slates in the middle of the table. Later on, in the midst of experiments with the single slate, he lifted the top slate of these two, placed some coloured chalks on the lower one, and placed the other slate again on top. The other important movements I may give in the words of my friend, Mr. J. Russell, who saw them. He had not been initiated into the *modus operandi*, but he did know definitely and positively that Mr. Davey's performances were not "mediumistic," and he was acquainted with the object of Mr. Davey's investigations. Mr. Russell is naturally an exceptionally keen observer, and he noticed and recorded the important trick-movements, of which there is no mention in the reports of the other two uninitiated witnesses. I therefore add his testimony to my own. Mr. Russell writes in his report:—

In the meantime, Mr. Davey had once more examined the two slates where the coloured chalks were, but finding nothing, had placed them side by side, and carelessly, as if in a fit of absent-mindedness, had taken the chalks from the slate which had been at the bottom, and placed them on the other. He had then put them together as before, except that the original position of the slates was reversed, the old bottom one being now at the top, and the old top one at the bottom. Presently, asking Mr. Padshah if in a former sitting with Eglinton the medium had not got some writing on his shoulder, he took up the two slates and placed them on Mr. Padshah's shoulder, but in less than a minute took them off, reversing them as he did so, and replaced them on the table. The old bottom slate was now once more at the bottom, and the old top one at the top, but each slate had been reversed, so that the two sides which had originally been turned to the table were now turned up. In a few minutes Mr. Davey had a sort of convulsion, Mr. Hodgson and Mr. Hughes said they heard sounds like writing, the slates were opened, and there, on the lower one, was a message, half in green, half in red (nearly the colours chosen by Mr. Padshah and my wife), expressing a hope that we should be satisfied with writing given thus, under such excellent test conditions.

Now, from the point of view of the psychologist analysing the value of human testimony, I regard Mr. Padshah's reports as in several respects the most instructive of the whole series. Mr. Padshah's mind is pre-eminently clear and sincere, and his report, written immediately after the sitting, is an excellent expression of the effects produced upon him. We can see, so to speak, a piece of wonderful testimony (as regards this particular manifestation) *in the act of making, but not made*. Describing the commencement of the sitting, Mr. Padshah wrote:—

There was full light on every corner of the table; two of my (?) slates, one washed by myself, the other by Mr. Davey, were put very nearly in the centre with a number of small chalk-pieces between them of different colours.

Later on, in discussing this manifestation, Mr. Padshah wrote:—

I confess I do not remember, even after such a brief lapse of time, whether I had examined the two slates not washed by me, and found them unwritten. I imagine I must have, for otherwise it would be very stupid.

Here, in the first place, we see that while Mr. Padshah's memory told him at the commencement of his report that one of these two slates had been washed by himself, his memory told him, apparently, a short time later that neither of these slates had been washed by himself. This, at least, seems to be the fair inference from his words. But the next point, concerning the examination of the slates, is more important. He imagines, he says, that he must have examined the slates, "for otherwise it would be very stupid." In the case of ninety-nine out of a hundred bona fide witnesses the statement in their report would not have taken this form. Their imagination that they must have examined the slates would have usurped the place of their failing remembrance, and they would have written, with perfect sincerity, "I examined the two slates and found them perfectly clean."

The same general method was employed in Sittings V., VI., VII., IX., XII., and XV., the slates being Mr. Davey's, and the communications having been prepared beforehand, and I think that the reader will have no difficulty now in supplying the omissions which vitiate the records. The choice of colours and the transcription of passages from books chosen by the sitters, and the writing in foreign languages, I shall consider later.

In Sitting VIII. the word *Yes* was found on the upper surface of the lower of two of the sitters' slates held together. This word was written on the top of one of the sitter's slates while the sitter was glancing over the books on the shelves for the purpose of choosing one for the locked slate experiment, I saw Mr. Davey write it on the slate, as the slate lay on the middle of the table, and then turn the slate over. This slate afterwards reached the required position by the regular method.

After this experiment came that of the locked slate, which was also a success, and while the sitter was wondering over the locked-slate message, Mr. Davey took one of his slates into another room and covered one side of it with writing. This interval the sitter speaks of parenthetically as a "momentary absence." After Mr. Davey's return the two-slate experiment was conducted in the regular way, and was indeed completed; but Mr. Davey got nervous, shuffled the slates out of position again, and hardly knew himself what had become of the writing. In trying to make up for this false move he slipped again, the sitter noticed a shuffle of the slates, seized them, and discovered the writing "before its time."

As regards the "two-slate" incident which occurred in Sitting XVI., Mr. Davey informed me that before the sitting began, and while exchanging greetings in the ordinary manner, he undid the parcel containing the slates brought by Miss Symons, took out one of her slates, substituted for it a new slate brought by himself, and tied up the parcel again,—all this with his back to the parcel so that his movements might be concealed from Mrs. Sidgwick and Miss Symons. He then left the room, ostensibly to fetch his own slates, &c., from another room, and while absent, wrote on the slate belonging to Miss Symons. Miss Symons herself carried the parcel, now containing Mr. Davey's slate, to the séance table without, of course, noticing anything wrong. Her slate, with writing on it, was placed among Mr. Davey's own slates, and, when the time came, in the course of the two-slate experiment, was re-substituted openly, as described in detail by Mrs. Sidgwick in the following account, which was written the day after the experiment, and before she knew how and when the trick was done. (The footnotes are a later addition.)

Miss Symons had brought two common slates and a locked one. Mr. Davey had also slates with him of various shapes; one of them with round corners as Miss Symons's had, and some with square corners. Miss Symons's two slates were held together on the table and under the table by her and Mr. Davey.[1] Then one of Mr. Davey's square-cornered slates was substituted for one of them, then again

---

[1] This is, of course, a mistake. It was the two slates out of the parcel, but one of these was really Mr. Davey's.

removed, and the two round-cornered ones again held, on the ground that though it might be easier to get writing on Mr. Davey's slate, it would be more satisfactory to get it on Miss Symons's.[2] We waited a considerable time. Mr. Davey asked me to draw the curtains between the two rooms. Then we sat as before, the two slates on one another on the table, and our hands on them. The sound of writing was heard, and presently on looking between the two slates, one of them was found to be written on allover one side. I cannot remember every detail of what occurred, but the impression produced on my mind most distinctly was, that one of Miss Symons's slates had been written on all over one side, and that there had been no possible opportunity for Mr. Davey to have done this.

We now come to the experiments with the locked slate. In addition to the two similar locked slates of small size which I have already mentioned, Mr. Davey had some other similar locked slates of large size, of the Faber make. One of these is represented in *Proceedings*, Vol. IV., pp. 466, 467. My impression is that he had *three* similar slates of the large pattern. He most frequently, however, used the smaller size, described by Mr. H. W. S. in Sitting XI., as "composed of two ordinary pieces of slate, about six by four inches, mounted in ebony covers hinged on one side with two strong plated hinges, and closed in front, beyond the question of a doubt, with a Chatwood's patent lock." Let us call these two locked slates A and B, and suppose that A is the first locked slate exhibited. I shall now describe in detail the locked slate incident in Sitting II. Mrs. Y.'s account of this is as follows:

He gave me a locked slate of his own, which I thoroughly washed and locked myself, and put the key in my own pocket. We then joined hands, and Mr. Davey and my daughter placed one hand each on the slate as it was lying on top of the table. Different questions were asked, and we waited some time, but no response came. Mr. Davey seemed to me very much exhausted, and I urged him to desist from any further efforts. But he seemed loth to do this, and said he would rest a little while, and would then, perhaps, be able to go on. After a short time of conversation, the slates all the while being in full view and carefully watched by me, we

---

[2] Really Miss Symons's this time.

again tried it, under the same conditions as before, only that this time Mr. Davey requested us each to take a book at random from the shelves in the room, and mentally think of two numbers representing a page and a line, and he would see if he could reproduce it. This also failed of any result, and Mr. Davey said he feared he was too tired to produce anything, as he had been very much exhausted by a long and very successful séance the night before. We again begged him to desist, but after a short rest he insisted on another trial. The slates still remained all the time in full view on the table. Mr. Davey asked my daughter to choose another book, which she did at random, he having his back to her and standing at some distance while she did it. This book was at once tied up and sealed by one of the party, Mr. Davey never touching it from first to last. I then held it in my lap, while we joined hands as before, and Mr. Davey and my daughter each put one hand on the slate. Still nothing came. Then we changed positions, and I placed my hand on the slate instead of my daughter, giving her the book to hold. During this change she kept her hand on the slate until I had placed mine beside it, and the book was awaiting her on the opposite side of the table, my husband all the while holding Mr. Davey's other hand. I am confident that Mr. Davey could not possibly have manipulated the slate during this change, for it was in full sight all the while, and our hands were on it, and the book was tied and sealed on the opposite side of the table. A few minutes after this readjustment Mr. Davey seemed to have a sort of electric shock pass through him, the perspiration started out in great drops on his forehead, and the hand that was touching mine quivered as with a nervous spasm. At once we heard the pencil in the slate moving, and in a few moments Mr. Davey asked me to unlock the slate. My daughter took the key out of her pocket and handed it across the table to me, and I unlocked the slate, and found it covered on both the inner sides with writing. When read, this writing proved to be a sort of essay or exhortation on the subject of psychical research, with quotations from the book chosen intermingled throughout. I forgot to say that Mr. Davey had asked us all to choose in our minds two numbers under ten to represent a page and a line of the book, but had finally concentrated his thought on what my husband was thinking. In the writing there were quotations from every page we had any of us thought of, but not always the line; but in the case of my husband the line was correct, but not the page. He had thought of page 8, line 8. The line was quoted from page 3, and Mr. Davey said this confusion between 8 and 3 frequently occurred, because of the similarity of the numbers. This test seemed to me perfect. The slate was under my own eye on top of the table the whole time, and either my daughter's hand or my own was placed firmly upon it without the intermission of even a second. Moreover, we closed and opened it ourselves.

This sitting was held in my rooms in the evening. In the morning Mr. Davey came to my rooms, and re-arranged some of my books. He placed a series of greyish-white books (chiefly Cambridge University Calendars) on a shelf easy of access, and in the middle of them he put a volume of selections from Mrs. Browning's poems after first copying some phrases from it. This book had a blue binding with gilt lettering on the back. The communication afterwards found by the sitter em bodying these phrases from the book, I saw Mr. Davey then and there in the morning write in the locked slate. At the first trial of this experiment, the volume of Mrs. Browning was not chosen. No result, therefore, was obtained. At the next trial Miss Y. chose the required book. Now, Mrs. Y. states that Mr. Davey had his back to her daughter and was "standing at some distance while she did it." Mr. Y. also says, "My daughter, leaving him at the table, replaced on the shelves the book she had first taken down, and took at random a copy of Mrs. Browning's poems." Miss Y.'s own account of this part of the incident is also positive:—

We sat as before around the table, discussing the failure of the experiment. Finally Mr. Davey started up and said, "We must try it with one book alone. Will you choose one, Miss ___?" I supposed that he asked me to do it because my seat was nearest to the bookcase. I got up and went to the bookcase. Mr. Davey stood by the table with his back to me. That latter fact I feel as if I remember most distinctly. I mention it to show that I chose my book at random and was not influenced in my choice by him.

As a matter of fact Mr. Davey escorted Miss Y. up to the book case and led her, as it were, up to the very shelf where the required book, in bright contrast to its dingy neighbours, was "forcing" itself to be chosen. "Choose a book, any book, take any book at random,"—with a wave of the hand in front of the special shelf, and Miss Y., quite naturally, reached out her hand and took the book that "fixed" her gaze. I gave in my previous notes what I thought was the probable explanation of the agreement of all the witnesses in the erroneous statement that Miss Y. went alone to the bookcase to choose her book. After the writing had been produced in the locked slate, Mr. Y. asked Miss Y. if she had gone alone to the bookcase, and she replied that she had, and that Mr. Davey had remained by the table with his back towards her. I conjectured also that Miss Y.'s lapse of memory was an instance of transposition, that she remembered correctly Mr. Davey's actions, but connected them wrongly with her second choice of a book instead of with her first.

By the "forcing" of this book the first step in the trick was performed. The next step was to substitute locked slate B for A. But the sitters were very careful, as the reader may notice from the accounts of their actions when Mr. Davey suggested that Miss Y. should change places with her mother. Miss Y. kept her hand upon the slate as she walked round the table, and correctly says that she did not relax her hold of the slate till her mother had her hand upon it. Nevertheless the time came when Mr. Davey did substitute B for A. But there is no mention whatever, in Mrs. Y.'s report, of the circumstances which enabled Mr. Davey to perform the substitution. Nor is there any mention whatever of these circumstances in Mr. Y.'s report. They are mentioned, however, in the report of Miss Y.

> Mr. Davey asked us each to think of two numbers as before. Finally he asked us to write them down on a slate. I wrote mine on one of our own slates so that he could not possibly see what I had written, and I placed it on the table away from Mr. Davey, and leaned my elbow on it. I think the others did the same with the other slates. To my remembrance, some of us watched the locked slate all the time while we were writing.

Miss Y.'s remembrance, about which she was apparently not sure, is not correct. At this juncture all the sitters forgot the locked slate and left it unguarded on the table. My impression is that all the sitters left the table, Mr. Davey having so candidly remarked that they must not let him see the numbers they wrote, and not let him even see the movements of the end of the pencil. I then saw Mr. Davey with the help of his duster (see p. 163) substitute locked slate B for A. I may mention that Mr. Davey gave a plausible reason for desiring the sitters to write their numbers down, viz., that a previous sitter had forgotten the numbers which he had finally chosen, and therefore could not tell whether the passages quoted in the writing were according to the chosen numbers or not.

In Sittings I., III., V., VIII., IX., and XV. the *modus operandi* will now be obvious. In each case the communication was prepared beforehand, and an opportunity was given for the substitution of B for A.

In connection with Sitting IV. it is noteworthy that Mr. Padshah, who was not perfectly satisfied that he had taken due precautions in examining the two single slates, and in seeing that all the surfaces were clean, did become absolutely convinced that the locked-slate writing, if not produced by chemical means, was "undoubtedly genuine." Mr. Davey, as he had done for Sitting II., came to my rooms in the morn ing, and placed in a "forcing" position, with the neutrally tinted numbers of the periodical *Mind* in its neighbourhood, Bastian's volume on *The Brain as an Organ of Mind*, a bright red book of the International Scientific Series. Mr. Davey wrote on the locked slate in my presence the communication afterwards found there by Mr. Padshah, including the words, "The Brain an Organ of Mind." When asked to choose a book Mr. Padshah finally chose (mentally) the periodical *Mind*, after having thought both of *The Brain as an Organ of Mind* and of *International Law*. Mr. Padshah's conclusion about this experiment was that it is "evident that Mr. Davey must have minutely studied the time it takes for complete precipitation; or that the whole precipitation takes place simultaneously; or that the phenomenon is undoubtedly genuine. The theory of writing without a chemical and then bamboozling me would be really contemptible." As I pointed out in my contemporary notes to this sitting, Mr. Padshah did nevertheless lose perception of the slate A for a short time, and during this interval Mr. Davey substituted B.

In Sitting VI. a double substitution was made. The sitter wrote a question in A. Mr. Davey substituted B, opened A, read the question and answered it, and re-substituted it again. Mr. Davey's usual method in these cases was to take A out of the room for the purpose of reading and answering the question. Later on, the substitution was made again for another experiment, B having been prepared beforehand, and the book to be chosen by the sitter placed in a "forcing" position.

Sitting X. was with a Japanese gentleman, and the locked slate used was of the large size. The first locked-slate experiment involved merely a simple substitution. This was all that was involved in the second locked-slate experiment also. The Japanese part of the message was easily enough obtained. Mr. Davey had met the sitter before and had obtained some information about him. He then went to the "Japanese Village" on exhibition in London, and for a consideration procured the services of an interpreter in translating and writing in Japanese on the locked slate the communication which Mr. Davey provided in English. The sitter says: "Once more I locked the double-slate... and put the key in my pocket and even sealed it myself." Mr. Davey suggested the sealing, but he substituted the second locked slate for the first before the sealing took place. (Compare Zöllner's experiment with Slade. Mrs. Sidgwick supposes that Slade substituted for two slates put together by Zöllner two other slates upon which he—Slade—had just written. *Journal* S.P.R., December, 1886, p. 481. This case of Mr. Davey's is exactly parallel.)

In Sitting XI. there was a double substitution in experiment [a]. In experiment [c] there was a single substitution. Experiment [d] is described by the sitter as follows:—

Lastly, as requested by Mr. Davey, I took a coin from my pocket without looking at it, placed it in an envelope and sealed it up. I am certain that neither Mr. Davey nor myself knew anything about the coin. I then placed it in the book-slate together with a piece of pencil, closed it as previously and deposited it on the table; and having placed my hands with those of Mr. Davey on the upper surface of the slate, waited a short time. I then unlocked the slate as requested, and to my intense amazement I found the date of the coin written, by the side of the envelope containing it.

The seal and envelope (which I have now) remained intact.

I do not recall with certainty what the coin was. Let us suppose it was a shilling. Mr. Davey beforehand wrote the date of a shilling of his own in locked-slate A, placed this shilling in an envelope and sealed it up, and placed this envelope also in locked-slate A, which at the beginning of the experiment he had concealed about his person. He then requested the sitter to take a shilling from his pocket without looking at it, to place it in an envelope and seal it up, place it in the locked slate (B), &c. The sitting was at Mr. Davey's house, and Mr. Davey provided the envelope, from the same packet, of course, as the one already containing Mr. Davey's shilling in locked-slate A. The sitter was requested not to look at his coin, ostensibly, I believe, on the ground of precluding thought-transference, but really so that the sitter might not know the difference between his own coin and Mr. Davey's. It is now plain that all the dexterity required in this experiment was a simple substitution.

In the locked-slate experiment described in Sitting XII. there was a double substitution. In the first locked-slate experiment in Sitting XIII. there was a double substitution. For the second there was a single substitution.

Mrs. Sidgwick has furnished the following account of the locked slate experiment in Sitting XV.: —

We then [after the writing of the word Melbourne] again sat at the table, Miss Symons next to Mr. Davey. She now took charge of the locked slate, which at this period was examined and was blank. [Then follows the account of obtaining the message with the Spanish sentence in it.] It was not easy to read, and while we were engaged in deciphering it Mr. Davey was still gasping and suffering apparently from the effects of the effort. He wandered restlessly about the room, with convulsive movements, &c. After a time he seemed better, and we determined to try another experiment. A book was chosen out of Mr. Podmore's bookshelf and laid on the table under our hands, and Miss Symons and Mr. Davey sat next each other, this time holding the locked slate.

The locked slate was unguarded while we were poring over the first long message, and there was plenty of time and opportunity then either to substitute another similar one or to write the message. Moreover, the book chosen was the one wished for by Mr. Davey. He made various objections and suggestions till I perceived that for some reason he wanted that one and chose it. I tried at first to choose a small book because I wanted the trick to succeed, and fancied it would be done by holding the book on the slate under the table and opening it there. Whether I should have been conscious of acting on anything but my unaided impulse [in choosing the large book] if I had not wanted to help Mr. Davey [by choosing a small one] I do not know.

I shall now describe the method of producing writing on the interior surfaces of common slates screwed and corded together and the knots of the cords sealed. For accounts of this experiment see Sittings XIII. and XIV. I quote here the account given in Sitting XIII.

I now took the two new slates which I had purchased, and which had never for a moment passed out of my possession, I even taking the precaution of sitting on them during the foregoing proceedings. I placed a piece of red crayon therein, and screwed them down top and bottom so tightly that by no possibility could even the thin edge of a penknife be introduced. I then corded the slates twice across and across, sealing them in two places with red and blue wax (for, of course, any attempt to remove the seals by heat would cause the colours to fuse, and thus immediately detect the artifice), stamping them with my own private signet. Mr. Davey placed the slates under the table, and requested me to name some word I would like written. I stipulated for "April." After a few minutes, during which I most carefully watched him, he returned them, and after 10 minutes' work, so tightly were they closed, I found exactly what I had desired.

... After perusal of above, considering that the expression, "I found exactly what I desired," might be liable to a possible misconstruction, I think it better to add that I state in the most unequivocal, explicit, and emphatic manner, that after Mr. Davey had returned me my two slates, secured as above described, and which I most carefully and minutely examined to detect any signs of tampering, finding, however, my seals intact and the cording and screws in exactly the same condition as when they left my possession a few moments before, and that the word "April," which I had asked for, was legibly written with the crayon, on one of the inside surfaces. Whether the top or bottom I did not observe. The apparently impossible having thus been solved as I hereby testify.

The sitter might also have sealed the screw-heads without preventing the performance of the trick.[3] Mr. Davey takes the slates thus

---

[3] In this connection the following extract from an account of a séance by Mr. T.O. Roberts, whom Mr. Davey characterises as "without exception the keenest witness I have ever met," may be of interest. The séance took place on April 23rd, 1887, later than any recorded in Mr. Davey's paper in Proceedings, Part IV. Mr. Roberts was, I believe, aware that Mr. Davey was a conjurer.

Mr. Roberts "purchased two common slates with wooden frames (8in. × 5in.) and rounded corners." He continues:—

I cleaned the slates myself and placed a small piece of grey chalk... between the slates, which I then placed together, noting which were the inner surfaces by a printed heading at the top of each; I next drilled six holes through the frames, one at each end, and two at either side, into which I drove six screws, these tightly binding the two slates together, placed my seal on the head of each screw, then bound the slates with thick cord and sealed the ends after tying the final knot

When I handed the slates, thus prepared, to Mr. Davey, he told me that the test was too severe, and that he did not think that it would be possible to produce the writing under such circumstances, but expressed his willingness to try.

Operations commenced by his placing the slates under the flap or leaf of the table near the corner, supported by the fingers of his right hand while his thumb rested on the table; with my right hand I held his left above the table and with my left I assisted in supporting the weight of the slates in the same manner as adopted by him.

The word selected by me to appear between the slates was "Parnell."

After remaining in this position for some fifteen minutes, during which time I watched his hand most carefully, and thwarted what appeared to me to be his several devices for diverting my attention, he informed me *he could not produce the writing unless I allowed him to take the slates out of the room*!

To this I assented, feeling that I was beginning to expose his inability to rival the "spirit-mediums" if only ordinary watchfulness were exercised. While these and similar thoughts crossed my mind, the door opened, he returned with the slates, having only been absent from the room 3 minutes. I then examined the slates most carefully, and I solemnly assert that my seals were intact in every case and that the slates were bolted together so tightly that it would have been impossible to introduce even the blade of a penknife between them, while my cord round them was as tight as when it left my hands, and the sealed ends were undisturbed.

The task of unscrewing the slates, &c., occupied several minutes, and this I performed myself, when I confess, greatly to my surprise, that the word "Parnell" was clearly and distinctly written on the inner surface of the lower slate. This I was at a loss to account for, especially as the piece of chalk that was enclosed had no sign of friction whatever upon it, this being evident at a glance, as the ends thereof had been newly broken.

prepared and places them in a horizontal position between his right leg and the adjoining leg of the table. He holds them in that position by the pressure of his right leg. He then takes from his hip pocket a wedge with a fairly sharp edge for insertion, but with the other edges smoothened so as to avoid indenting the frames of the slates. I think that the wedge that Mr. Davey used was made of brass, and was somewhat more than two inches long and about half an inch wide. He forces this wedge between the frames of the two slates at a point farthest from the screws. Thus if the screws are on the top and bottom of the slates, he forces the wedge in at the middle of one of the sides. There is enough elasticity in the frames and the cords to prevent any injury to the frames or the cords or the seals. An opening a quarter of an inch wide is easily produced in this way. Leaving the wedge in position he takes from the hip of his trousers, where it has been fixed by the insertion of its ends in two small rubber loops, a piece of an umbrella rod, say seven or eight inches long, in the end of which is fastened a piece of pencil or chalk. This he inserts through the aperture produced by the wedge, and writes the words required. He withdraws the rod and the wedge, replaces them in their private receptacles, and brings the slate above the table.

The writing or drawing produced under an inverted tumbler placed on a slate on top of the table is described in several accounts. (Sittings I., II., and XVI.) The following is Mrs. Y.'s account of this experiment in Sitting II.:—

He placed one of our slates on three little china salt-cellars that lifted it up about an inch from the table. Upon the middle of this he placed several pieces of different coloured chalks, and covered them with a tumbler. Then he told my husband to form a mental picture of some figure he wished to have drawn on the slate under the glass, and to name aloud the colour he would have it drawn in. He thought of a cross, and chose aloud the blue colour. I suggested that blue was too dark to be easily seen, and asked him to take white, which he agreed to. We sat holding hands and watching the pieces of chalk under the tumbler. No one was touching the slate this time, not even Mr. Davey. In a few minutes Mr. Dav-

---

I neither know nor pretend to understand how this trick is done, but I congratulate Mr. Davey on the celerity displayed by him, and the skill he undoubtedly possesses.

ey was again violently agitated as with an electric shock, which went through him from head to foot, and immediately afterwards we saw, with our own eyes, each one of us, the pieces of chalk under the glass begin to move slowly, and apparently to walk of their own accord across the space of the slate under the tumbler. My husband had said just before that if the piece of red chalk under that tumbler moved, he would give his head to anyone who wanted it, so sure was he that it could not possibly move. The first piece of chalk that began to walk about was that very red piece! Then the blue and white moved simultaneously, as though uncertain which was the one desired. It was utterly astounding to all of us to see these pieces of chalk thus walking about under the glass with no visible agency to move them! All the while Mr. Davey, whose hands were held on one side by myself and on the other side by my husband, seemed to be on a great nervous strain, with hot hands and great beads of perspiration. When the chalks stopped moving, we lifted the tumbler, and there was a cross, partly blue and partly white, and a long red line marking the path taken by the red chalk! We were impressed by this test beyond the power of words to declare. The test conditions were perfect, and the whole thing took place under our eyes on top of the table with no hands of anybody near the slate.

The ostensible reason for placing the slate on the salt-cellars was that the slate might be insulated, so that the explanation of "electricity" might not be offered! Mr. Davey has a fine silk thread attached to one end of a button on his waistcoat. To the other end is fixed a small piece of *red wax*, which except when in use in the experiment is in his pocket together with the slack of the silk thread. While placing the slate on the salt-cellars with his left hand he takes the piece of wax between the fingers of his right hand, picks up with these same fingers some pieces of chalk,—moves his right hand forward to the other side of the slate—not yet placed in position—so that the thread shall be *under* the slate when the slate is placed on the salt cellars. He then places the slate in position, brings his hand down from the far side of the slate and places the pieces of chalk and the piece of wax on the middle of the slate, and places the inverted tumbler over them. But while he is making a little heap of the chalks on the middle of the slate, before placing the tumbler in position, he also draws a figure (or a number, as the case may be) that he thinks the sitter is most likely to choose. He draws this, of course, very rapidly and dexterously, and he arranges the chalks over it so as to conceal it. Further, he has placed the piece of wax on the side of the heap which is nearest to himself. He now takes his place very carefully so that the thread, the length of which has been well calculated, shall not be tightened too soon. The reader will now see that by with drawing his body from the table, Mr. Davey can finally cause the wax to move in the opposite direction, *i.e.*, away from himself, and through, so to speak, the little group of chalk fragments, producing a movement in them also. The tumbler is then lifted in excitement, usually by Mr. Davey, the slate is inspected, and a figure discovered. In the meantime Mr. Davey gives a jerk to the thread, moves away from the table, and gathers the wax and thread once more into his pocket.

There are several minor details of Mr. Davey's performances which hardly need explaining. Thus many of the sitters describe the pieces of pencil or chalk as being worn at the conclusion of an experiment. Usually they would have been found equally worn at the beginning of the experiment had the attention of the sitters been then called to them. Sometimes, indeed, they were not worn at the beginning, but Mr. Davey then took care to substitute worn pieces before the writing was produced. There are several specific cases (Sittings I., IX., and XIII.) where the pencil was found resting at the end of the message. These were in locked-slate experiments. Mr. Davey had chosen and placed the pencil so that when the slate (to be substituted) was closed, the pencil did not move when the slate was shaken. When the slate was carefully opened, right side up, the pencil was found where Mr. Davey had placed it.

After the foregoing explanations I believe that the reader will find little difficulty in explaining to himself Mr. Davey's modus operandi in most of the experiments in the series of sittings with him recorded in Vol. IV. of our *Proceedings*. But I shall give the details of a few other cases where either possibly the reader may still be unable to see the exact method used, or where a special additional trick was involved.

In Sitting III. occurs the following description:—

The next experiment was the placing of 3 bits of coloured chalk on the table, and of a clean slate (selected and placed by myself) over them. I put my hand on the slate, Davey his on mine, and we joined contact. Again we heard the sound of writing, and when I lifted the slate there was written large and neatly in the coloured chalks (three lines or so in each colour) this message:—"Don't you think I've done enough for you tonight I'm tired Joey." I noticed the chalks seemed worn, showing signs of work, just like the little bit of pencil in the previous experiment.

Mr. Davey very rarely used the "trick slate," but the case described above was one instance of its use. The slate was neither selected nor placed by the sitter. Mr. Davey first placed some coloured nibs of chalk on the table just in front of the sitter. He then took one of his own slates which the sitter had not touched, and apparently sponged both sides thoroughly. Mr. Davey himself then placed the slate over the pieces of chalk, and asked the sitter to place his hand upon the slate. The sitter then for the first time touched the slate. The slate used in this case had a false flap, which fitted the frame. On the surface of the slate itself, under this flap, was a prepared communication. The exterior surface of the flap resembled the exterior surface of the true slate. The interior surface of the flap was covered with a piece of blotting-paper which bore marks of use in the form of casual blots and lines, &c. On the table in close juxtaposition lay a pile of blotting-paper, the top of which was also marked by casual blots, &c. This blotting-paper was there for the ostensible purpose of drying the slates. Mr. Davey took the slate with false flap uppermost, and sponged the exterior surface of the flap. He then turned the slate over on top of the blotting-paper, and sponged the other surface of the slate. He then lifted the slate and placed it above the chalks, leaving the false flap behind upon the pile of blotting-paper, but with the blotting-paper side of it uppermost. The sitter was watching the slate "with all his eyes," but, of course, saw nothing which it was undesirable that he should see. "The chalks seemed worn," as the sitter says, but then the sitter did not examine the chalks beforehand, or he would have found them equally worn then.

There are two reports of Sitting VII., and I may refer to an experiment where the reports differ in a very important point. Mr. V.'s report is: —

The medium tore off half a sheet of letter-paper bearing the address of his house; this he gummed to the surface of an ordinary slate, a fragment of lead pencil was put on the paper, and the slate then transferred beneath the table-flap, and held by Mr. P. and the medium. Writing immediately audible. At our request the slate was exposed before it had ceased. To the best of my remembrance the slate could not have been beneath the table-flap for more than 20 seconds. On examination we found the following message written in a hand which bore a much greater resemblance to the medium's than any of the others.

Mr. M. writes:

Mr. Pinnock asked if we could not get the writing on a piece of paper instead of the slate. Mr. Davey said we might try, and thereupon tore a sheet of writing-paper into two, and pasted one half on to a slate by the four corners; he cut off a small piece of black lead from the end of a pencil, put it on the paper and covered the slate with another slate. Writing was heard at once, and we separated the slates and found the paper written over diagonally as in the case of the first slate. The paper was not, however, quite full, and it looked as if the slates were separated too soon, as the sentence was not finished. The writing was evidently written with the point of the pencil.

This experiment was actually "led up to" by Mr. Davey, who had already prepared the message, and who substituted the slate containing the prepared message by the two-single-slates method already described (p. 162). Mr. Davey also suggested that the slates should be examined before the sound as of writing had ceased.

In her report of Sitting XVI. Miss Symons describes one experiment as follows:—

He took up 12 squares of paper, asked me to name any 12 animals I liked, whose names he wrote on the 12 squares of paper. These were shuffled together, and I was asked to choose one, which I was to glance at and then instantly to burn. Mr. Davey at the same time threw the other squares into the fire. I next wrote the first and last letters of the animal I had chosen on another piece of paper, this Mr. Davey burned in the gas, bared his arm and showed us that there was nothing written there, rubbed the ashes of the burnt paper over the bare arm, and presently what looked like letters became very faintly visible. They did not, however, become sufficiently distinct to enable us to read them, and Mr. Davey said he would presently get the animal's name written on a slate.... Before he left, Mr. Davey held a slate with me under the table, and asked that the name of the animal written on the slip of paper I had chosen should be written on the slate. Writing was heard, the slate brought up, and I found "rhinoceros" wrongly spelt—in red chalk. This was correct, though how Mr. Davey knew, or by what means the word was written, I have no idea, for the slate appeared to me to be clean when we put it under the table.

This, though not on this occasion completely successful, is a very easy trick. Before the experiment, write on the arm, with a brush or a feather, in uric acid, the name of an animal (or a flower, or a country, &c.) likely to be one of twelve chosen. Wait till it dries. There is then no visible trace upon the arm. When the sitter names an animal, write, on the square of paper, the name that you have written upon your arm. Do the same with every piece of paper, no matter what animal the sitter names. The slip chosen afterwards by the sitter will necessarily contain the name written upon your arm. Rub the ashes of this paper upon the arm, and the letters will "stand out" in the colour of the ash. I have performed this experiment myself successfully two or three hours after writing the name upon my arm. The word "rhinoceros" was already written on the slate when Mr. Davey placed it under the table, as Mrs. Sidgwick had good ground for stating, for Mr. Davey wrote it openly in her presence and showed it to her while Miss Symons was out of the room.

Those who have read thus far, and who have taken pains to compare my explanations not only with those accounts which I have re-quoted in this article, but with the reports as originally given in Vol. IV. of our *Proceedings*, will realise now, I trust, if they have not done so previously, the extreme imperfection of those reports, and therefore the great unreliability of any testimony to the ordinary "slate-writing" performances of professional mediums. The medium may leave the room, he may withdraw a slate several times, he may separate slates placed together on the table, and alter the irrespective positions, he may turn slates over together, and yet not one of these circumstances may appear in the report of the sitter. These points and others the student might easily have discovered for himself by comparing the different reports given of the same sittings by the uninitiated witnesses, and yet these points are all of the most fundamental importance as regards the question of trickery. Thus in Sitting II. only one witness out of three refers to Mr. Davey's leaving the room, and only one witness out of three mentions the withdrawals of the slate before the writing was manifest. In Sitting IV. only one witness out of three records the separation of slates placed together on the table, &c. Further, in Sitting II. only one witness out of three records a highly important incident (the sitters' writing down on slates the numbers of which they were thinking), which was specially brought about by Mr. Davey for the express purpose of making a substitution, and during which the substitution was actually made.

Yet here I must confess that while it is gratifying to learn that Mr. Davey's labours have been so successful in producing the conviction that his "manifestations" and those of certain professional mediums do actually belong to the same category, it is disappointing to find that the chief object of at least my own Introduction to Mr. Davey's investigation seems to have met with but little appreciation by Mr. Wallace and those whom he represents. I admitted that "there are numerous records of 'psychographic' phenomena that have occurred with mediums (and also with Mr. Davey) which, as *described*, are inexplicable by trickery," and I endeavoured to show "how far such records might be misdescriptions, and what were the chief causes of the misdescriptions." The notes to the records were made for the purpose of showing to investigators some of the important misdescriptions that actually occurred, and that are therefore to be expected in such records. Further, there were five sittings each of which was reported by more than one witness, and opportunity was given to the student to discover for himself, by a comparison of different reports of the same sitting, numerous other instances of misdescription. The question of primary importance concerns the value of human testimony under the circumstances involved. Why do we not accept such testimony? Because it is demonstrably fallible in precisely those particular points where it must be shown infallible before the phenomena can be accepted as super normal. I have already briefly adverted to some of the instances of this fallibility in the explanations which I have given of Mr. Davey's methods, but it seems to me needful to further emphasise it in view of the fact that Mr. Wallace has been able to entertain the idea that Mr. Davey was a "medium." My purpose will, I think, be sufficiently conserved if I refer to one or two additional striking cases of discrepancies between reports of the same sittings.

In Sitting I. a long message was obtained on the locked slate, but the message was incomplete, ending "We hope to." Mr. Davey ended the message purposely in this way and afterwards "led up to" the request that the message might be concluded. In the meantime Mr. Davey had written the conclusion of the message on one of Mr. R.'s slates, which was lying on the table, writing downwards, ready for the experiment.

Mr. R. writes as follows:—

I desired after this to have the writing on the double slate of Mr. Davey's continued at the point where it had been broken off, and obtained this result on one of my slates which I held underneath the table and which began immediately. "We hope to see you again—Joey." I was also anxious to know what the VII signified as I have already said before. On the first attempt we got the answer—"good-bye Joey"—but we were more successful on again putting the question, the result being a distinct "Septe—"; whether, as I have already said, it was intended for September I cannot tell.

Mr. L.'s account is:—

The writing having stopped so abruptly, two ordinary slates were placed upon the table in the manner before described, and it was asked by Mr. R. that the letter should be concluded. Within a period of 15 seconds from the time of asking such question and after completing the circle with our hands, the words "to see you again, Joey," were written.

The two slates were again placed in the same position as before, and Mr. R. having put an unimportant question, after the completion of the circle as before, I saw upon the slate "Good-bye, Joey"; but on a second trial a scrawl was obtained which looked very much like "Sept. Joey," but it was impossible to say definitely what it was intended for.

It is noteworthy that Mr. L. makes this experiment follow immediately after the locked-slate message, and places the "tumbler" trick last, while Mr. R. makes it follow the "tumbler" trick, which he puts immediately after the locked-slate experiment. I do not recall what Mr. Davey told me about his precise operations in connection with these writings, but from my knowledge of his methods, aided by his note, I infer that he cleaned the top of the slate, the underside of which was already prepared with the conclusion of the message, that he placed this slate under the table, turning it over in the process according to the single-slate method described on p. 256. He then wrote "Good bye, Joey," on the then under surface of the slate with his thimble pencil, brought the slate up and laid it upon the table, when the words "hope to see you again, Joey," were manifest. These words were rubbed out, and this slate placed upon another slate, and both slates together placed under the table, being reversed in the process. He then wrote "Sept. Joey" on the under surface of the bottom slate, brought both slates together to the top of the table, and lifted the top slate, when the words "Good-bye, Joey," appeared on the upper surface of the bottom slate. He rubbed these out, put this slate upon the other, and placed them once more under the table, reversing them as he did so, and then possibly, as though changing his mind, placed them on top of the table again. When the top slate was removed, "Sept. Joey" appeared on the upper surface of the bottom slate.

But I wish to draw the reader's particular attention here, for a reason which will appear later, to the fact that one witness states that the communication came upon a single slate held underneath the table (a statement which Mr. Davey confirms), and the other that it came between two slates placed on top of the table. Nor is this the only instance of a mistake of this kind in the reports. Comparing the accounts of the letter-paper incident which occurred in Sitting VII., and which I have quoted on p. 277, it will be noticed that in one account the experiment is described as having been made with a single slate held underneath the table, in the other as having been made with two slates above the table. The experiment was actually made with two slates which were probably finally held under the table.

Another important discrepancy between the reports of Sitting VII. occurs in the case of the ordinary two-slates experiment. Mr. V., after referring to the locked slate and the writing of a question therein, &c., describes experiments [a] and [b], and then proceeds to describe experiment [c] as follows:—

Two ordinary slates taken, cleaned by us, but not marked, pieces of red and green chalk introduced between them, the slates then deposited in front of the medium in full view, and about four or five inches from the edge of the table and from the medium's body; the medium rested one of his hands on the upper surface of the top slate, and my hand reposed on his.

After a pause the sound of writing distinctly audible; this continued for about 15 seconds, then the medium remarked, "What a pity I forgot to ask you what colour you would have it in." Mr. M. suggested green; sound of writing continued for about five seconds longer, then ceased. On the removal of the top slate, the bottom slate was found to be completely covered with writing. The writing ran in diagonal lines across the slate; the writing was upside down with respect to the medium; the writing was firm and distinct in character. The first three-quarters of the message were written in red, the last quarter in green.

Mr. M. is much more accurate in his account of this incident, and I include, in the quotation which I give, his reference to other experiments which came between the beginning and the end of the two-slates experiment. There is no clue to the *modus operandi*, for the uninitiated reader, in Mr. V.'s account, but there are very obvious clues, for any careful student of the series of reports, in Mr. M.'s account. Mr. M. describes the locked slate, &c., &c., and then proceeds:—

Mr. Davey then showed me some ordinary slates, in wooden frames. These I helped him to wash and dry. We then took our seats round the table. Mr. Davey asked Mr. Pinnock to place the locked slate under his (Mr. Pinnock's) coat and then button up the coat.

[c] We now took three slates, on one of them we placed three fragments of crayon, two of which were red, the other green, we then covered up this slate with another and left them on the table in full view.

[a] On the third slate we also put a piece of crayon and then held the slate underneath one flap of the table which we put up for the purpose. ... We sat in this way talking and smoking for some time, twenty minutes half an hour I should say, nothing whatever occurring. At last Mr. Davey asked me to change places with Mr. Pinnock. This I did and thus had one

of my hands on the slate. Mr. Davey now said, that in the manner usual at séances we would ask questions of an imaginary being; and he said, "Are you going to do anything tonight, Joey?" After a short pause he repeated the question, and then I felt the slate vibrate as if being written on, and could hear a scratching noise; we took the slate from under the table-flap and saw the word "yes" written over Mr. V.'s initials, and I particularly noticed that the writing was towards Mr. Davey, and upside down to him, and in all we saw afterwards this was the case.

[*b*] I now asked a question as to the whereabout of a person at that time, not knowing the answer myself; we waited for some time without any result, when Mr. Davey asked me to again change places with Mr. Pinnock.

[*d, e, g,* & *c.*] I did so, and Mr. Davey told Mr. Pinnock to place the locked slate on the table beside the two slates we had left face to face, and we also lifted the uppermost of these two slates and found the slates still quite clean, with the three pieces of crayon between them. We again waited some time with no results; meantime, having a discussion as to mediumship of different people, and then Mr. Davey asked if I were a medium. After a pause I heard vigorous scratchings on the two slates left face to face on the table and on which Mr. Davey's arm was resting, his two hands being engaged, one in holding the slate under the table flap, the other in holding Mr. V.'s hand; the scratching lasted roughly under ten seconds, and I expected to see a dozen words or so, and was therefore amazed to discover, when the top slate was lifted, that the underneath slate was covered with writing from corner to corner, and also the writing was not straight across the slate, but was across it diagonally; three-quarters of the writing was in red, the other quarter in green, and *no crayon was left*.

Now, the reader will easily infer from Mr. M.'s account that the two slates were placed in position by Mr. Davey and were wrongly *supposed* by the sitters to have been taken from those cleaned by them some time—not immediately—previously. After studying the accounts of Sittings IV., especially the reasoning by Mr. Padshah, and after considering that the slates used were Mr. Davey's, the reader will also infer that the under surface of the bottom slate was already covered with the writing afterwards found. He will then argue that if a series of movements such as those described by Mr. Russell (see p. 264) could be completely omitted from the report of Mr. Padshah, they might also have occurred in Sitting VII., although they are not recorded in the reports of that sitting. But more instructive even than these clues in Mr. M.'s account to Mr. Davey's *modus operandi*, is the fact that Mr. V. describes the steps of the experiment as though they came in immediate sequence; whereas we learn from Mr. M. that about half an hour elapsed between the first and last steps of the experiment, and that during this interval experiments were being made with a single slate. These experiments [a] and [b] are described by Mr. V., but they are described as occurring before the commencement of [c].

I shall give one more illustration of differences between reports of the same sitting before proceeding to consider in detail those (slate writing) sittings which Mr. Wallace has mentioned as being, apparently, particularly hard to explain. In his report of Sitting VII. Mr. V. writes:—

At the request of the medium, Mr. P. wrote a question in the book-slate (I shall call this slate A in future); he then locked it and pocketed the key. Neither Mr. M. nor I knew the nature of the question at the time. The slate was left for some minutes upon the seat of an arm-chair, but was subsequently transferred first to Mr. P.'s coat, and then to the table at which we sat.

Later on, after recounting experiments with a single slate and with two slates, Mr. V. continues:—

The medium and Mr. P. placed their hands upon slate A, which had remained in sight in front of the latter since the commencement of the séance. The sound of writing audible almost immediately. Mr. P. opened slate, and we found the question he had written, together with the accompanying answer.

Turning now to Mr. M.'s account, we find that he also mentions that Mr. Pinnock, at the beginning of the sitting, wrote a question in the slate, locked it and kept the key. He says nothing, however, as to what was done with the slate at that time, but goes on to describe, as the next events, the examination of the table, the cleaning of ordinary slates, seating themselves at the table, *&c.* He then writes: "Mr. Davey asked Mr. Pinnock to place the locked slate under his (Mr. Pinnock's) coat and then button up the coat." Then follows his description of the two-slates experiment, which I have quoted above (p. 282), on reference to which it will be seen that Mr. Pinnock placed the locked slate on the table before any writing had been obtained between the two single slates. After the writing between the two slates was obtained, the locked-slate experiment was proceeded with.

Mr. Davey now put his hand on the locked slates which had been left on the table since Mr. Pinnock took them from under his coat; we heard scratching inside.

Putting these accounts together, it is obvious where the opportunities for substitution were given. B might have been substituted for A shortly after A was locked up and while it was resting on the arm chair as described by Mr. V., so that it was really B that Mr. P. placed in his pocket (a desirable place lest the sitter should think of examining it before A was re-substituted). The re-substitution of A was easy while the sitters were absorbed in the long message that appeared between the two slates.

Compare, with these reports, those of Mr. R. and Mr. L. of the locked-slate experiment in Sitting I. The first mention of this experiment by Mr. L. occurs after the experiments with the single slate and with the two slates together. He writes:—

Mr. Davey then produced a "locked slate," which I examined *most minutely*, and, as far as I was able to judge, the surfaces were genuine slate and had not undergone any process of preparation which would aid him in obtaining writing. A small crumb of pencil was inserted, and the slate closed and locked by Mr. R. The key was then given into my possession. We then placed our hands in an exactly similar position as before, and Mr. R. having repeated the question, "Will the Emperor of Germany live through the year?" I very soon heard the pencil travelling over the surface of the slate. After the lapse of about four minutes the slate was carefully unlocked by Mr. R., and the pencil very much worn was found at the place where the writing ended.

From this account it would seem that the first inspection of the locked slate almost immediately preceded the production of the writing, but it appears from Mr. R.'s account that the slate was inspected and locked at the very beginning of the sitting, and was put by him in the pocket of his coat. After describing the experiments with a single slate and with double slates, he continues:—

The next experiment was with Mr. Davey's closed slate. After it had been produced from my pocket we laid it on the table locked and with the small piece of pencil inside, joined hands as before and the question was put, "Will the Emperor of Germany live through the present year?" Immediately the writing began, exactly the same as on previous occasions, and when after the space of 4 minutes (about) I carefully unlocked the slate we found the following wonderful message.

If the reader will compare these accounts with the accounts of the locked-slate experiment in Sitting VII., and especially with that by Mr. M., he will at once surmise that the locked slate was produced from Mr. R.'s pocket before the communication was obtained between the two slates in the preceding experiment, and that while the sitters were absorbed in its contemplation Mr. Davey substituted B for A. But the important point to notice is not how the trick was done. The important point is that just as we have seen from the reports of Sitting VII. that a witness may describe the steps of the *two-slates experiment* as though they occurred in immediate sequence, with no other experiments intervening, whereas in reality the last steps were separated from the first by an interval of half an hour, during which other experiments were made with a single ordinary slate, and the locked slate also claimed attention:—so here we find, from the reports of Sitting I., that a witness may describe the steps of the *locked-slate experiment* as occurring in immediate sequence, with no other experiments intervening, whereas in reality the last steps were separated from the first by an interval during which various experiments were made with a single slate and with two slates together.

Bearing in mind, then, these two special possibilities of error and also the other possibilities of error to which I have drawn attention on pp. 260, 263, 269, *all of which are sufficiently demonstrated by comparing the reports of the sitters themselves*, let us now consider in detail the reports of Sittings XI. and XII., which Mr. Wallace has particularly mentioned (*Journal* S.P.R., March, 1891) as needing explanation.

The report of the experiments in Sitting XI. is as follows:

[a] After I had finished examining the [locked] slate, Mr. Davey asked

me to write in the slate any question I liked while he was absent from the room. Picking up a piece of grey crayon, I wrote the following question: "What is the specific gravity of platinum?" and then having locked the slate and retained the key, I placed the former on the table and the latter in my pocket.

After the lapse of a few minutes I heard a distinct sound as of writing, and on being requested to unlock the slate I there discovered to iny great surprise the answer of my question: "We don't know the specific gravity, JOEY." The pencil with which it was written was a little piece which we had enclosed, and which would just rattle between the sides of the folded slate.

Having had my hands on the slate above the table, I can certify that the slate was not touched or tampered with during the time the writing was going on.

[*b*] Next; having taken an ordinary scholar's slate and placed a fragment of red crayon upon it, Mr. Davey placed it under the flap of the table. I held one side with my hand as before. I then heard the same sound as previously, and when the slate was placed on the table I found the following short address distinctly written: "Dear Mr. S——, The substitution dodge is good; the chemical is better, but you see by the writing the spirits know a trick worth two of that. This medium is honest, and I am the only true JOEY." The writing was in red crayon, and was in regular parallel straight lines.

[*c*] Then, again, Mr. Davey requested me to place a small fragment of slate pencil in the lock slate, which latter had been previously cleansed with sponge by me. Respecting the method of closing the slate, &c., everything was done as in the first instance; the slate was locked, and I retained the key.

As soon as the sound of writing was over I picked the slate from off the table, where it had been lying right under my eyes, unlocked it, and read as follows: "We are very pleased to be able to give you this writing under these conditions, because with your special knowledge upon the subject you can negative the theory of antecedent preparation of this slate as advanced by certain wiseacres to explain the mystery.—'JOEY.'" The fact that the pencil when removed from the interior of the slate had diminished in size and showed distinct traces of friction convinces me that it was the pencil and nothing else which produced the caligraphy. If the particles taken from the pencil by friction did not go on the surface of the slate, where could they go?

Mr. Davey's Imitations by Conjuring, &c.

[*d*] Lastly, as requested by Mr. Davey, I took a coin from my pocket without looking at it, placed it in an envelope and sealed it up. I am certain that neither Mr. Davey nor myself knew anything about the coin. I then placed it in the book-slate together with a piece of pencil, closed it as previously and deposited it on the table; and having placed my hands with those of Mr. Davey on the upper surface of the slate, waited a short time. I then unlocked the slate as requested, and to my intense amazement I found the date of the coin written, by the side of the envelope containing it.

The seal and envelope (which I have now) remained intact.

This last feat astonished me more than the others, so utterly impossible and abnormal did it appear to me. I may also mention that everything which was used, including the cloth and sponge with which the slates were cleansed, were eagerly and thoroughly scrutinised by me, and I failed to detect anything in the shape of mechanism of any kind.

Now, that this report is very scanty and inadequate is obvious on the face of it, and Mr. Davey assured me that there were other experiments tried of which no mention appears in the report. But I do not propose to depend, for my explanations of this sitting, simply and merely upon either my remembrance of conversations with Mr. Davey or my detailed knowledge of his methods. I am anxious that students should learn how to interpret for themselves such accounts as these, and it seems to me that I can best achieve this result by pointing out, to begin with, some of the most obvious indications, which we find in the report itself, of its deficiencies, afterwards amending the report as regards its most flagrant misdescriptions. We shall then easily see how the tricks were performed.

In the first place, then, let us note that various important circumstances receive no mention whatever in the report. Mr. S. tells us that after he locked the slate in experiment [*a*] he placed it on the table, and "after the lapse of a few minutes" he heard "a distinct sound as of writing," &c. But he tells us absolutely nothing as to what happened during this interval which he describes as a "few minutes." He does not even mention the return of Mr. Davey to the room. The locked slate might have been changed a hundred times for all that appears to the contrary in the sitter's account. What he certifies is "that the slate was not touched or tampered with *during the time the* [sound as of] *writing was going on.*" The reader may also notice that the sitter does not say anything about enclosing a piece of pencil when he first locked the slate, but it appears afterwards that "the pencil with which it [the answer] was written was a little piece which we had enclosed." Here is another indication of circumstances omitted. When did "we" enclose it?

Concerning experiment [*b*] Mr. S. writes: "Having taken an ordinary scholar's slate and placed a fragment of red crayon on it, Mr. Davey placed it under the flap of the table. I held one side with my hand as before." *Before? when?* The sitter makes no mention of any previous experiment where he assisted in holding a slate under the table, yet his remark here carries a clear implication that there was at least one such previous experiment. Again, the sitter says that *Mr. Davey* placed the slate under the table, and he does not say that any examination of it was made by himself. There is, therefore, nothing in his description of this experiment which conflicts with the supposition that Mr. Davey took a slate with the writing already on one side, slipped it under the table, turned it over, pressed it against the flap, and then asked the sitter to join in holding it against the table.

Similarly, in his account of experiment [*c*] there is no express statement that conflicts with the supposition that the slate might have been changed during the interval between the sitter's examination of the slate and the beginning of the sound as of writing. The sitter says nothing as to the interval that elapsed between the time of his depositing the slate, after locking it, on the table, and the conclusion of the sound as of writing, except the remark, "where it had been lying right under my eyes," and he does not expressly say that it had been "lying right under his eyes" during the whole of the interval in question. The inference from his remark, comparing it with his account of experiment [*a*], is that what he meant was that the slate had been "lying right under his eyes" during "the time the [sound as of] writing was going on." Other experiments occupying, say, half an hour might have been in progress between the time of the sitter's locking the slate and hearing the sound as of writing.

Now, we have already seen that witnesses may make numerous positive and express statements which are entirely erroneous, the result being that *if their descriptions are taken as correct*, the phenomena which they describe are inexplicable by trickery. But, curiously enough, in the report before us, the phenomena described in the three first experiments mentioned by the sitter are perfectly explicable by trickery without altering a single word of his accounts of them. Only in his account of experiment [*d*], and scarcely in that, are the details narrated in such a way that, *as described*, trickery seems impossible.

Let us now revise the report of these four experiments. We may do this by consideration of the methods usually adopted by Mr. Davey as revealed in the whole series of sittings, and by consideration also of the errors to which a witness is liable, as revealed by a comparison of the different reports given of the same sitting. We find, then, that Mr. Davey usually began by giving the locked slate to the sitter to examine, and possibly to write a question there in. He then tried experiments with a single slate and with two slates together, and afterwards recurred to the locked slate. Observing this general order, I amend the report as follows, correcting, of course, by no means all of its fundamental misdescriptions; but—and I desire to lay very special emphasis on this fact—the changes which I do make, excepting the descriptions of the actual substitution of one slate for another, and the doings of Mr. Davey while out of the room, *are all warranted by a comparison of the reports of those sittings where more than one independent report was made*. The additions which I make are in square brackets, and the italicised parts explain how the tricks were done.

[*a*] After I had finished examining the slate, Mr. Davey asked me to write in the slate any question I liked while he was absent from the room. Picking up a piece of grey crayon, I wrote the following question: "What is the specific gravity of platinum?" and then having locked the slate and retained the key, I placed the former on the table and the latter in my pocket.

[When Mr. Davey returned to the room, he asked me to examine the table carefully, which I did. It was an ordinary table, without any trick mechanism of any sort. *During this interval Mr. Davey substituted B for A*. He then gave me some ordinary slates to wash and dry. *During this interval Mr. Davey left the room, opened A and answered the question, and returned and re-substituted A for B*. Mr. Davey now took three slates. [*b*] On one of them (*which had not been in the hands of the sitter, and on the under surface of which was the prepared message*) he placed a fragment of red crayon. He then covered up this slate with another and left them on the table in full view. On the third slate he also put a fragment of crayon and held it under the table against the flap and asked me to hold it on my side. We asked if we should get any phenomena, and after a short time the sound of writing was heard, and we looked at the slate and found the answer "Yes." We then put our hands on the locked slate.]

[*a*] After the lapse of a few minutes I heard a distinct sound as of writing, and on being requested to unlock the slate I there discovered to my great surprise the answer of my question: "We don't know the specific gravity, Joey." The pencil with which it was written was a little piece which we had enclosed, and which would just rattle between the sides of the folded slate.

Having had my hands on the slate above the table, I can certify that the slate was not touched or tampered with during the time the writing was going on.

[*c*] Then, again, Mr. Davey requested me to place a small fragment of slate-pencil in the lock slate, which latter had been previously cleansed with sponge by me. Respecting the method of closing the slate, &c., everything was done as in the first instance; the slate was locked, and I retained the key.

[In the meantime Mr. Davey lifted the top slate of the two on the table, but there was no writing there. *He reversed the positions of the two slates so that the slate with the message on the under surface was now on top.* Mr. Davey then took these two slates and placed them under the flap of the table, *reversing them together as he did so.*]

[*b*] I held one side with my hand as before. I then heard the same sound as previously, and when the slate was placed on the table I found the following short address distinctly written: "Dear Mr. S——, The substitution dodge is good; the chemical is better, but you see by the writing the spirits know a trick worth two of that. This medium is honest, and I am the only true Joey." The writing was in red crayon, and was in regular parallel straight lines.

[*While the sitter was examining this message Mr. Davey substituted B for A.* Mr. Davey now put his hands on the locked slate. Very soon the sound of writing began.]

[*c*] As soon as the sound of writing was over, I picked the slate from off the table, where it had been lying right under my eyes, unlocked it, and read as follows: "We are very pleased to be able to give you this

writing under these conditions, because with your special knowledge upon the subject you can negative the theory of antecedent preparation of this slate as advanced by certain wiseacres to explain the mystery.—'Joey.'" The fact that the pencil when removed from the interior of the slate had diminished in size and showed distinct traces of friction convinces me that it was the pencil and nothing else which produced the caligraphy. If the particles taken from the pencil by friction did not go on the surface of the slate, where could they go?

[*While the sitter was copying the communication Mr. Davey left the room and placed the coin in envelope, and envelope in slate A, and wrote date* (see p. 271) *and returned.*]

[*d*] Lastly, as requested by Mr. Davey, I took a coin from my pocket without looking at it, placed it in an envelope and sealed it up. I am certain that neither Mr. Davey nor myself knew anything about the coin. I then placed it in the book-slate together with a piece of pencil, closed it as previously and deposited it on the table.

[Mr. D. showed and explained to me a means commonly employed in producing slate-writing by fraud. *While the sitter was examining the trick slate, Mr. Davey substituted A for B.* I then took the locked slate.]

And having placed my hands with those of Mr. Davey on the upper surface of the slate, waited a short time. I then unlocked the slate as requested, and to my intense amazement I found the date of the coin written, by the side of th eenvelope containing it.

The reader now will surely need no further enlightenment as to the details of events in Sitting XII., or indeed in any other sitting of the series, and will, I trust, be disposed to think, with Mrs. Sidgwick and myself, that the most startling result of Mr. Davey's investigation is not the wonder of the tricks themselves, but the extreme unreliability of the accounts given of them by uninitiated witnesses. And we should remember further that these accounts probably represent the most accurate reports, as a whole, of such performances, ever brought together in a series. For the witnesses knew beforehand that they were expected to write out accounts of what occurred, and more important still, the reports were written within two or three days after the sittings. And I may here also refer the reader to my remarks in *Proceedings* S.P.R., Vol. IV., pp. 396-399, concerning the disadvantages under which Mr. Davey laboured as compared with the ordinary professional medium.

The reader may now ask how far his knowledge of Mr. Davey's methods may prevent him from being imposed upon by fraudulent mediums. Possibly not very much. Frequent observation and practice of them, however, would no doubt be of great assistance. Surreptitious writing on slates held under the table, the substitution, openly made, in the case of slates lying on the table,and the manipulations of two slates where the writing is originally on the under surface of the lower slate, and eventually is found on the upper surface of the lower slate, are, I have no doubt, in frequent use by fraudulent mediums. Eglinton was apparently in the habit of using all these methods. Mr. Davey purchased some of his devices from an individual who gave him to understand that they had been procured from an American medium. The author of *Revelations of a Spirit-Medium* enumerates eight different methods of apparently producing "independent slate-writing" without the help of a confederate. The most important of these is a variation of the two-slates experiment combined with the trick-reading of pellets, and as I have reports in my possession written by a member of our Society who witnessed this general method in the case of two well known American mediums, Watkins and W. A. Mansfield, I quote what the author of the book says (pp. 124-126) about this trick.

Another feat that is astonishing and convincing is accomplished with two clean slates. They are thoroughly cleaned and laid side by side upon a table, on one side of which sits the "sitter," and opposite him the "medium." The "sitter" is now furnished with a small square of soft white paper and requested to write the name of some deceased friend or relative, and with it a question. This being done he is requested to fold it up small, similar to the physician's powder papers. The "medium" has a blank one, folded in the same way and palmed between the index and middle finger of the right hand. When the "sitter" has folded his pellet, the "medium" reaches forth his right hand and takes it between the thumb and index finger and carries it to his forehead. While raising the hand to the head, he slips the written pellet down and the blank one in view. After holding it to his forehead a few seconds he requests the "sitter" to take it and hold it against his own forehead for a moment. Of course the "sitter" gets the blank pellet and the "medium's" hand drops to his lap. He now opens the pellet and reads it. We will say it reads: "John Smith. Will my business succeed? George."

Having read it and palmed it again, he now requests to hold the pellet to his forehead again. He effects the change and says to the "sitter"; "You now hold the pellet in your left hand and I will write the answer."

This time the "sitter" has the pellet he wrote, and holds it while the "medium" takes up a slate, and leaning well back, holds the slate with his left hand and body, and writes with the right hand in such a position that the "sitter" cannot see the writing. He writes:

"Dear George,— Your business is sure to succeed beyond your expectations. John Smith."

He now states to the "sitter" that he does not feel at all sure that he has written the correct answer, and reads aloud:

"The papers will never be found. Harry White."

Of course it is not an answer to the question, and the "sitter" so states. The "medium" requests that he open the pellet and see if it is plainly written, with no omission of words.

While he is doing so the slate is deftly turned the other side up. When the sitter reports that the question is properly and plainly written, the "medium" *apparently* rubs off the line of writing and lays the slate on the table, writing underneath. He now announces that he will let the spirits do their own writing, and putting the other slate on top of the one containing the writing lays his hands on top of the slate a few seconds, when he opens them, and of course there is no writing.

He now states that he does not believe he can get anything—but, wait, he says, we will put the pellet inside—*that* may help them.

The pellet is placed on the blank slate and the one containing the writing laid on top. Now the writing is between the slates. In picking up the two slates together, he turns them over, and the writing is on the bottom slate. He now allows the "sitter" to hold the slates alone, and indicates when to open them. They are opened, and much astonishment created by the pointed answer to the question inside the pellet.

It is obvious that there may be many variations of this trick. In the sittings with Watkins and Mansfield, accounts of which were written by Mr. John F. Brown, an Associate Member of the American Branch, each of the sitters (three at one sitting and two at the other) wrote several questions on slips of paper afterwards crumpled or folded into pellets. Some of the medium's surreptitious dealings with their pellets were observed by the sitters. Mr. Brown's account of the two slates incident at the sitting with Watkins is as follows:—

Watkins "gave the name of George Hall, and soon commenced to write rapidly, covering one side of the slate, then he turned the slate over on his arm so that the writing could not be seen, and wrote a few lines more. He said we had better copy the messages as it would be more interesting for us to have them to refer to. A. took pencil and paper, and Watkins read slowly the following communication. ... The side of the slate containing the signature was turned towards us without any concealment; the opposite side was kept from our view. After he had finished reading, and while we were looking at the copy, Watkins erased the part we had seen, then turned the slate end for end, rubbed the sponge again over the same side and put the slate on the table with writing on its under side. Not long after the George Hall message, a second attempt was made to get independent writing, a first attempt having been unsuccessful. The previous attempt was shortly before the first message, and its lack of success gave Watkins the excuse for writing himself. A bit of pencil was now laid on the top of a clean slate and the slate with the writing already on it lifted from the table and placed upon the other. Watkins then took hold of them both, waved them in the air, and, as he brought them back, turned the mover so that the slate now under neath had writing on the upper side. All this was distinctly followed by us both, and we were looking for writing just where it appeared."

The writing that appeared, as Mr. Brown points out, was doubtless what Watkins had written when he was pretending to write the first part of the George Hall message; and when he pretended to read the George Hall message from the slate he "made it up" as he proceeded until he turned the slate over.

I witnessed yet other slight variations at a sitting with a Mrs. Gillett. Under pretence of "magnetising" the pellets prepared by the sitter, or folding them more tightly, she substitutes a pellet of her own for one of the sitter's. Reading the sitter's pellet below the table, she writes the answer on one of her own slates, a pile of which, out of the sitter's view, she keeps on a chair by her side. She then takes a second slate, places it on the table, and sponges and dries both sides, after which she takes the first slate, and turning the side upon which she has written towards herself, rubs it in several places with a dry cloth or the ends of her fingers as though cleaning it. She then places it, writing downward, on the other slate on the table, and sponges and dries the upper surface of it. She then pretends to take one of the pellets on the table and put it between the two slates. What she does, however, is to bring the pellet up from below the table, take another of the sitter's pellets on the table into her hand, and place the pellet which she has brought up from below the table between the slates, keeping in her hand the pellet just taken from the top of the table. The final step is to place a rubberband round both slates, in doing which she turns both slates over together. She professes to get the writing without the use of any chalk or pencil. Some of her slates are prepared beforehand with messages or drawings. More interesting, perhaps, because of its boldness, is her method of producing writing on the sitter's own slates. Under pretence of "magnetising" these she cleans them several times, rubs them with her hands, stands them up on end together, and while they are in this position between herself and the sitter she writes with one hand on the slate-side nearest to herself, holding the slates erect with the other hand. Later on, she lays both slates together flat on the table again, the writing being on the under most surface. She then sponges the upper surface of the top slate, turns it over, and sponges its other surface. She next withdraws the bottom slate, places it on top and sponges its top surface, keeping its under surface carefully concealed. The final step, the reversal, is made, as in the other case, with the help of the rubber band. Mrs. Gillett has probably other methods also. Those which I have described were all that I witnessed at my single sitting with her.

In many records which have been written of experiences with "pellet mediums," the writers affirm that the medium never touched their pellets. In the case of such records we are fully justified in applying our general conclusions, drawn from a consideration of the errors made by sitters with Mr. Davey, although Mr. Davey did not, at any of the sittings reported, use the "pellet" device. If a *bona fide* witness can report with confidence that he held his hands on the slate and watched it continuously during the experiment, when in reality he completely forgot about it for an appreciable interval during which it was manipulated by the "medium," he can equally report that he watched his pellets the whole time, and that they were not touched or tampered with by the medium, although as a matter of fact the medium did touch them, and did substitute one pellet for another. A good instance of this has been brought to my recollection by the following memorandum, which I have just found among my notes:—

October 8th, 1888.

On Saturday morning, October 6th, Mrs. [Q.] called, and during the conversation referred to the medium Watkin and a conversation we had had concerning him the previous week. She had then been profoundly influenced by sittings which she had with him, and had been most strenuous in denying that Watkins touched the pellets in any way, although admitting that Watkins had rolled up one piece of paper as specimen, and left it on the table, and she was unable to say what had afterwards become of it. She had also been positive that Watkins did not tamper with the slates which she was holding.

On Saturday morning Mrs. [Q.] told me that since leaving me the previous week she had recalled that twice, at least, Watkins had touched the pellets, once when he moved one pellet aside, saying "this is mine," and on another occasion when he took up a pellet and asked her to pinch it up a little smaller.

Lapse of memory again, we must note, rather than mal-observation. Similarly, Mr. Padshah originally scouted the suggestion that he had lost perception of the locked slate, but when I assured him positively that he had lost perception of it, he was finally able to discover, in a dim recess of memory, on its way to oblivion, the occasion of the loss. And Mr. Padshah's report was written immediately after the sitting. When we reflect on circumstances like these, how manifestly absurd appears the reliance which so many Spiritualists place upon reports of "psychographic" and kindred phenomena, where the lapse from memory of possibly a single apparently trivial detail vitiates the whole record of the uninitiated witness.

Slade also uses the two-slates method, as appears from the following account of a sitting which I had with him in February, 1891:—

February 10th, 1891.

Sitting with Slade, 11 a.m., February 3rd, at 229, East 14th-street,

Mr. Davey's Imitations by Conjuring, &c.

New York, with Mr. Z.

Second room—simple table with two leaves—large Pembroke. On further side of table was a small table close to the large table and close to the wall, with a cloth over it hanging down. [The accompanying rough diagram will illustrate the positions.]

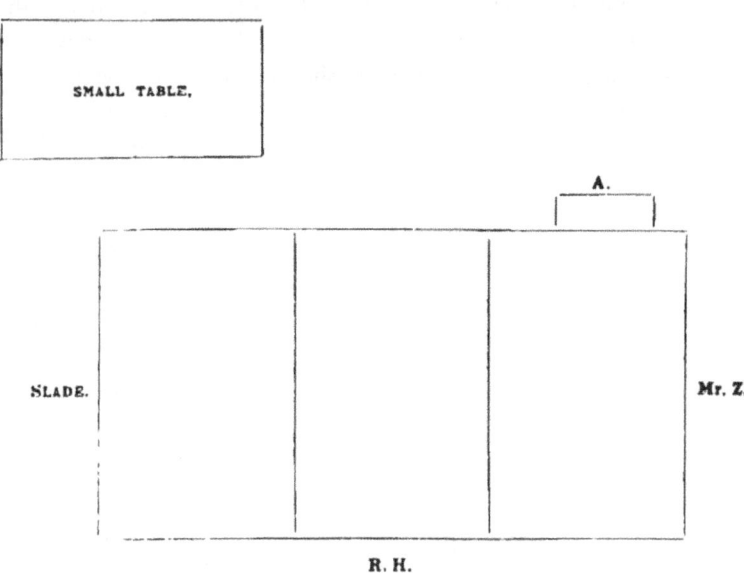

Slade took two slates from the little table, upon which there were four, and showed them to us. He turned them over and let us turn them over they were clean. He then replaced them on the small table, and suggested trying to get "raps." He asked if Mr. Davis was present—three raps. Just before this he sat somewhat facing me, with right leg visible and left leg partly so, saying, "Notice my position." Almost immediately, however, he turned square towards the table and his left leg disappeared from view entirely. The raps, which continued for some time—two or three minutes—might easily have been produced by his foot. All our hands were together upon the table.

Slade then took one of the replaced slates from the side table, put a piece of pencil on it (he had a box full of small fragments on the table), and held it with his right hand under the table for a short time. He did not keep the slate close to the table. No result. He then replaced the slate on the table, and then took from the side table one of the two slates which we had not inspected (the surface of which was much newer looking), and placed it on top of the slate No. 1. Then seizing both slates together with his left hand, he turned them over and laid them on the right arm of Mr. Z., slanting away from me. Almost directly the sound as of writing was heard. Slade's fingers were concealed behind the slate, but I observed the tendons working in his wrist.

The message was an ordinary general statement signed by the name Davis. The writing, of course, was already on the under surface of slate No. 3 when Slade took it from the side table.

Then Slade took the clean slate and held it under the table—then said, "They're taking it away from me," and stooped and pressed up against the table as though his arm was being drawn under. Half of the slate then became visible [at A] facing me to the right, then disappeared, and shortly after struck me at the lower part of my waistcoat and fell to the floor. All this might easily have been done by Slade with his feet.

He placed the slate on the table again and turned it over. It was clean. Then he took it in his left hand, and stooped down somewhat to the left. I then heard a sound as of a slate slipping to the floor, and conjectured from the position of the upper part of his left arm that he was substituting the slate which we had just seen to be clean for slate No. 5, which was probably out of sight and leaning against the leg of the little table. (See *The Seybert Commission on Spiritualism*, p. 74.) In making the substitution I suppose that one of the slates slipped on the floor. However, he brought this slate (No. 5) to the top of the table, took No. 2 or 4 (I am not sure which) and placed No. 5 on top of it. Then seizing both slates together with his right hand, he turned them over and laid them on my left arm slanting away from me and away from Mr. Z. Almost immediately the sound as of writing began but the slates slid along my arm slightly, bringing Slade's fingers in gentle contact with my arm, and I could feel the motion of a finger or fingers moving backwards and forwards. Slade also noticed the contact, and drew the slates up further so that his hand did not touch my arm.

The writing was a general kind of message, signed by T. Z.—the name of Mr. Z.'s father, but as Mr. Z., at a sitting ten days before, had been specially asked by Slade what his father's name was, there was no test. The two writings were evidently by the same hand.

After this, Slade asked me to write a question on the side of a slate (No. 2 or 4) remote from himself. I wrote, "Fred, will you give me a test if you are here?"

Slade took the slate in his right hand with the question on the under surface and held it under the table, not close to the leaf. Almost immediately the slate rubbed past my left knee, suggesting that Slade was turning the slate over. I then observed him furtively looking downwards, and he shortly asked me if I had asked two questions. I replied, No. After waiting a little longer, he said that he felt that there was no influence present the power, he thought, was exhausted, and he could generally tell when it left him. He thought it was no use sitting any longer. He suggested that I should have another sitting soon, alone.

(Mr. Z. thought the writing between the slates remarkable, and had no idea whatever of the trick movements, &c., made by Slade. I explained the details to him immediately after we left.)

There remained to ascertain the truth of my conjecture concerning slate No. 5. I requested permission to thoroughly examine the large table, and began by turning it completely over to my right, so that I could see the corner where I supposed the fifth slate to be. As I did this, Slade carelessly stooped down, picked up the fifth slate from the floor, close to the foot of the small table, and laid it by the other slates on the table.

Further, I have proof that Mr. Davey's general methods are easily discoverable from the reports themselves by persons who have paid special attention to the production of such phenomena by trickery. About a year ago I became acquainted with a Mr. W. S. Davis, of New York, a printer by profession, who was making himself familiar with the methods used by fraudulent mediums in rope-tying, slate-writing, materialisation, and other "physical phenomena." I requested him to read the accounts of the sittings with Mr. Davey and write me a description of the methods which he supposed Mr. Davey used. His descriptions were practically correct throughout, and indeed he gave additional variations of some of the methods. The only cases where the reports failed to give him sufficient clues were the book incidents, where the communication was prepared beforehand and the book was "forced." Mr. Davis himself has given some sittings which have been regarded as specially remarkable by various Spiritualists of New York and Brooklyn, and brief accounts of these have appeared in some Spiritualistic papers. Mr. Davis informs me that he never claimed that he was assisted in any of his performances by "departed spirits," and as a matter of fact, they were all due to trickery, and he has explained to me his methods in detail. It may be interesting to compare the reports given by "Spiritualists" of a sitting with Mr. Davis with his account of what actually occurred. But I shall first give the explanation of Mr. Davey's "materialisation" séance which has been furnished by Mr. Munro, who assisted Mr. Davey, or rather, I should say, actually produced the phenomena.

The following is the report by Mr. R. of the sitting for materialisation:—

On Thursday evening, the 7th October, 1886, I was present at a séance held by Mr. Davey, at his house. There were in all eight persons, myself included. We took our seats at 7.30 p.m., round an ordinary dining-room table (in the dining-room of the house), which, at Mr. Davey's request, we examined carefully, as also any other objects in the room which demanded our attention. The door of the room was locked, and I placed the key in my pocket, it was also sealed with a slip of gummed paper; the gas was then turned out, so that we were left in darkness. A musical box was wound up, and set to play an air, with the object, as I suppose, to enliven the proceedings! I held Mr. Davey's right hand, his left was held by Mrs. [J.]; the rest joined hands, so that during the séance a continual chain was formed which was maintained the whole time. After we had remained some time thus, various noises as of a shuffling of feet, &c., were heard in different parts of the room, and I distinctly felt something grasp my right foot; almost immediately I was touched on the forehead by a cold hand, which, at Mr. Davey's request, also touched those that wished it. The musical box was lifted, and although it was dark I fancied I saw it, surrounded by a pale light, descend through the air; it certainly struck me lightly on the side of the head, then it was again raised, and deposited on the table.

The hand which touched me was cold and *clammy*; it evidently belonged to a most courteous and obliging spirit, for it did exactly what we desired! and at my wishing to feel the full palm on the back of my head (so as to ascertain its shape and size) it rested there for fully three seconds; it was, however, a somewhat weird experience! Various raps were now heard, a gong sounded behind my back, and we were told by Mr. Davey to pay attention, as something wonderful was about to take place. Faintly, but gradually growing more distinct, a bluish white light appeared hovering about our heads; it gradually developed more and more till at length we beheld what we were told was the head of a woman. This apparition was frightful in its ugliness, but so distinct that everyone could see it. The features were distinct, the cheek bones prominent, the nose aquiline, a kind of hood covered the head, and the whole resembled the head of a mummy. After favouring those of the company who wished to see its full face by turning towards them, it gradually vanished in our presence. The next spirit form was more wonderful still; a thin streak of light appeared behind Mr. Davey, vanished, appeared again in another part of the room, and by degrees developed into the figure of a man. The extremities were hidden in a kind of mist, but the arms, shoulders, and head were visible. The figure was that of an Oriental, a thick black beard covered his face, his head was surrounded by a turban; in his hands he carried a book which he occasionally held above his head, glancing now and then from under neath it. The face came once so near to me that it appeared to be only two feet from mine. I thus could examine it closely. The eyes were stony and fixed and never moved once. The complexion was not dusky, but very white; the expression was vacant and listless. After remaining in the room for a few seconds, or rather a minute, the apparition gradually rose, and appeared to pass clean through the ceiling, brushing it audibly as it passed through. The séance here terminated; the gas was turned on again, and everything appeared the same as when we first sat down; the door was unlocked, the seal being found intact. I will mention that during the whole of the séance I held Mr. Davey's right hand, with but one exception, when it was found necessary for him to light the gas to see to wind up the musical box, as it had stopped playing. *Nothing was prepared beforehand; the séance was quite casual;* we could have sat in any room we wished, and we had full liberty to examine everything in the room, even to the contents of Mr. Davey's pockets, which were emptied (before beginning the séance) by him on the table before our eyes!

October 8th, 1886.   JOHN H. R.

Mr. Munro's Account.

Although Mr. Davey was kind enough to instruct me in the methods of his slate-writing, I was not present at any of the sittings described in Vol. IV. of the *Proceedings*, with the single exception of the materialisation séance, which is the only one published in which confederacy was employed.

The explanation which I am about to give of that séance may be of interest as indicating how much or, I should rather say, how little, accounts of such phenomena correspond with the facts which actually occur. And the sitting for materialisation is eminently adapted for this purpose, inasmuch as the accounts were written so very soon after the sitting ended, two of them at least having been completed on the same evening. At the same time I should like to remind the reader that any explanation I can give of the phenomena can be but partial. I can only inform him of the *mere mechanical processes* which were employed. A full explanation would involve a description of the mental attitude of the sitters and of every word and gesture of the medium whereby that mental state was altered. *This* I cannot describe, and yet it is of infinitely greater importance than any of the tricks and devices used in the manufacture of the spirit forms. The latter might have been produced by a host of other methods, and the method actually employed does not appear to me to be a matter of much consequence. It is only for the reason stated above that I consider its publication of any value. With slate-writing séances it is quite different, for in them the mechanical processes—the mere "conjuring"—which can be used are necessarily very limited. But when we darken the room and keep the investigators in ignorance of what they are to observe, the possibilities for trickery are infinitely increased, whilst the control which the medium must exercise over the thoughts and emotions of his sitters need not be so great. It is not, therefore, surprising that Mr. Davey himself introduced the accounts of the sittings for materialisation with a sort of half apology, seeing that the testimony for such phenomena is and must be so much inferior to that for slate writing. To myself it is even surprising that any explanation should have been called for in the case of a séance where the facilities for deception were so great, and it is almost incredible that an investigator of Mr. Wallace's experience should regard it as of equal or even greater importance than the slate-writing experiments. Considering the sensational nature of the phenomena observed, it is not surprising that the accounts of this sitting show an even greater discrepancy with fact than do those of the slate-writing séances, and I think it will be well to indicate first of all a few of the more important errors.

In the first place, the séance, so far from being the casual affair which Mr. R. supposed, had, in fact, been carefully arranged beforehand. I had been staying with Mr. Davey for several days before the sitting, and we had discussed the details of the materialisation process, and even rehearsed it through, the night before it was given. Mr. Davey had also given a similar séance in the spring of the same year.

In the second place, the locking and sealing of the door, so carefully recorded in all three accounts, was by no means so well calculated to prevent the entrance of agencies from without as the reporters appear to have imagined. The process of "locking" the door, which was performed by Mr. Davey himself, although he subsequently gave the key to Mr. R., consisted in first locking and then unlocking it. Sealing a door with a piece of gummed paper is now a well-known trick. The gummed paper, if properly adjusted, adheres firmly to the door when it is opened, and, when it is again shut, presents all the appearances of never having been moved. The interesting part about the sealing in this case is that the paper was not properly adjusted, and at the end of the séance, Mr. Davey, noticing that the gummed paper had fallen down onto the ground, hastily stuck it back in its place and called Mr. R.'s attention to the fact that the door was still sealed—a fact to which he and the other sitters readily gave their testimony.

The third point to which I would call especial attention is the examination of the room, with which every one of the three reporters was quite satisfied, Mrs. J. even going so far as to state "we searched *every* article of furniture." In spite of this positive statement, the examination was imperfectly performed, for in that cupboard beneath the bookstand, which was situated furthest from the door, were concealed a gong and several other appliances, including the female spirit herself. Mr. Davey showed his sitters that the other cupboard was empty, but diverted their attention from this one so skilfully, that they were afterwards convinced that they had examined it also.

I will now describe what took place at the séance, step by step so far as I can remember it.

It had been arranged to hold the meeting in the dining-room, and Mr. R.'s statement, "we could have sat in any room we pleased," is not correct. Mr. Davey did, indeed, I believe, offer his sitters their choice of rooms. But had they selected any other room (and there were only two others which could conveniently have been used for the purpose) he could easily have found some excuse for rejecting it in favour of the dining-room. There was no peculiar advantage in this room. It was selected chiefly on account of its size, and because it was not overcrowded with furniture. At the same time the cupboard behind the medium's chair was conveniently situated. With regard to the sitters, four had been expected. The arrival of Mrs. J. and Miss W. was, as Mrs. J. remarks, quite unexpected. Mr. R. certainly intended to be present, but I am not sure whether he anticipated a séance. On their entrance into the dining-room, free leave was given the sitters to search every article of furniture, and I think the search was pretty well performed until it came to the cupboard under the bookshelf. From it Mr. Davey diverted attention by emptying out his pockets before his audience — a proceeding which they did not fail to remember as a conclusive proof of the completeness of their search.

Mr. Davey now "locked" the door in the manner already described and the gas was turned down. At the same moment I, who had by this time found my way into the passage, and could hear everything which was taking place in the room, turned down the gas outside, in order that no light might enter the room when the time for my own entry should arrive. A large musical box was then started, not, however, as Mr. R. supposed, "to enliven the proceedings," but that it might help with Mr. Davey's shuffling of feet to cover any noise which I might make in entering the room.

I must now explain that Mr. Davey sat at the end of the table with his back turned towards the bookshelf and with the door on his left. Mr. R. was on the right, Mrs. J. on the left of the "medium," the other sitters being seated on either side of the table nearer to the window. Having put out the gas in the passage I opened the door very slowly and came in barefooted, closing the door behind me as noiselessly as I could. In so doing I will not be certain that I was not responsible for

Mr. Davey's Imitations by Conjuring, &c.

one or two of the very conclusive spirit-raps mentioned in the reports. I now went up to behind Mr. Davey's chair, and, after tapping him on the back to indicate my safe arrival, proceeded to raise the musical box and wave it to and fro above the heads of the sitters, and to make raps in different parts of the room.

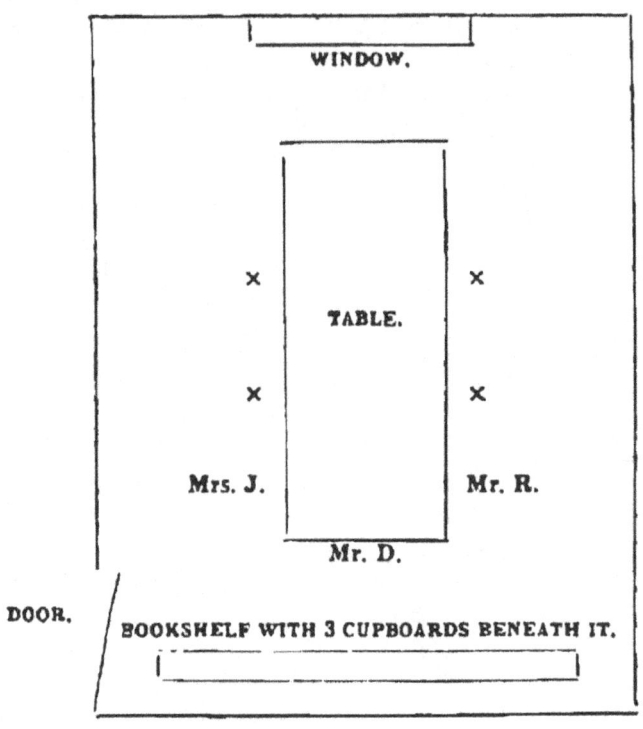

Throughout the séance I maintained a position behind Mr. Davey's chair, never advancing at any time in the direction of the window. The knocks at the far end of the room and on the ceiling were made with a long stick which I had brought in with me. Touching the sitters on the face, feet, or hands was of course easily managed, and, inasmuch as I had rolled up my sleeves and held my whole hand and forearm in a jug of cold water before coming in, Mr. R.'s description of the cold clammy hand which touched him was not purely imaginative.

I next opened the cupboard beneath the bookshelf behind and

to the left of the medium. It contained the gong, which I sounded for some minutes, and also the first spirit-form, which I afterwards divested of the black cloth with which it was draped. This spirit is graphically described by Mr. R. as "an apparition frightful in its ugliness, with cheekbones prominent and nose aquiline, the whole resembling the head of a mummy." It was prepared as follows:— A mask was taken and fixed upon a thick piece of cardboard. Muslin was arranged round the mask, and a thick collar of cardboard coated with luminous paint encircled the whole. The collar had been exposed to the sun throughout the day, so that when I uncovered the form it was rendered distinctly visible by the light thrown upon it by the now luminous collar. This spirit-face is interesting as indicating one method of producing materialisation phenomena without the aid of an accomplice, for a conjurer of Mr. Davey's skill would have had but little difficulty in manipulating it in my absence.

The second spirit was personated by myself. A turban was fixed upon my head, a theatrical beard covered my chin, muslin drapery hung about my shoulders. The book from which I read was a portfolio coated inside with luminous paint. It was concealed in the cupboard, where it lay wrapped up in black cloth, and when this covering was removed the book gaped a little. and so gave rise to the thin streak of light which Mr. R. describes. Before materialising, I mounted upon to the back part of Mr. Davey's chair, from which position I gained several advantages. At one moment I could bend forwards so as to appear close to the table in front of the medium, and at another, by standing upright, I could bring my head close to the ceiling. Indeed, the range of possible movement is so great, and the effects so startling, that many people have difficulty in believing the above explanation until they have seen the process repeated in a lighted room. My face and shoulders were rendered visible by the light thrown upon them by the open "book" which I was supposed to be reading, so that Mr. R. could not possibly have seen me when I held it above my head. For the "fixed and motionless" condition of my eyes I cannot account, the pallor of my face was due to flour, "the vacant and listless expression" is natural to me.

The statement that the apparition appeared to pass clean through the ceiling with a scraping noise occurs in all three reports. It is a curious mistake, founded on a blunder which I made in the acting of my part—a blunder so serious that at the time I thought I had—in part at any rate—betrayed the secret of our ghostly methods. When I had, still standing on Mr. Davey's chair, risen to my whole height, I gradually elevated the "open" book above my head, shut it and firmly pressed the two sides of the cover together. But the portfolio had been exposed to the sun all the day and the cover had in consequence become warped, so that its free margins were bent away from one another. When I pressed them together, they adhered for an instant and then burst as under with a loud report which was mistaken by the listeners for the brushing of the spirit form against the ceiling.

The séance did not terminate immediately after this, as Mr. R.'s account seems to suggest, but a very considerable interval elapsed, during which I slowly found my way out of the room. Mr. Davey then lit the gas in the dining-room, whilst I at the same time turned up the gas in the passage outside and then retreated upstairs—there to remain till the sitters should depart. I believe the statement that the medium's hands were held continuously throughout the séance except when he was turning on the musical box or lighting the gas is perfectly correct.

And now I think I have sufficiently explained the methods employed in this materialisation séance, and the reader has probably already long ago come to the conclusion that the sitters were in this case peculiarly unscientific and ill-suited for the investigation of these phenomena. To have neglected to lock the door themselves and yet to suppose it had been carefully locked, to have omitted to search in one of the cupboards and yet to imagine they had searched "every article of furniture," does indeed appear extraordinary neglect and carelessness. And yet I do not think there are many persons who would have taken these precautions, simple as they seem, or, if they had done so, would not have neglected other tests equally important. After all, their omitting to lock the door themselves was not an important error on the part of the sitters, since nothing would have been easier than for Mr. Davey to have provided me with a duplicate key, although I will not be sure whether he had two keys in this particular case.

I may also add that the cupboard, in which the gong and other materialisation apparatus were concealed, was only a small side cupboard wholly hidden from view by the door of the large central compartment when the latter was opened, and it needed a very careful investigation to discern that it did not form part of the central cupboard itself. The omissions, therefore, of the sitters are not so absurd as they at first appear. Let the reader also remember that Mr. Davey had given séances more or less similar to this one before other sitters, and yet in no case had any suspicion of his *modus operandi* entered their heads. I may further mention that in one case I had assisted in the production of the phenomena (which, however, did not in this case include materialisation) when the sitting was not held in the medium's own house.

For my own part, I think that the three published reports of this sitting are by no means extraordinary for their inaccuracy, and this conclusion is borne out by a reference to other reports of séances—reports by both sceptics and Spiritualists. Except in those rare cases in which I was myself present at the sitting, it is impossible for me to say where exactly the misstatements came in. But in most of the accounts which I have read there are grave omissions, in many there are positive statements about facts, which the writer could not possibly have known, and in almost all there is a confusion between the phenomena observed and the inferences drawn from those phenomena. And this is true even of the accounts written by trained scientific men—even of what I may call the classic literature of Spiritualism.

The errors of the reports appear even less striking when we take into account the fact that the accuracy of a report naturally varies inversely with the miraculous and sensational nature of the phenomena. If reports of slate-writing séances are full of mistakes, those of materialisation séances must be much more so. Indeed, I doubt whether the most scientific and unemotional of men would be capable of giving an accurate account of every detail they had seen after having been in the presence of the "figure frightful in its ugliness," or of that other form with the stony and fixed eyes which Mr. R. describes. But, even so, the results are extraordinary, unless one recognises their true explanation, which is to be found not in the inferiority of the sitters but in the superiority of the medium.

The methods devised by Mr. Davey were simple enough—so simple that one is astonished at his boldness in using them. But, as I have already pointed out, the mere mechanical means which he used were nothing, the personality of the medium was everything. Had silence been maintained throughout the séance, success would have been impossible. Had Mr. Davey and myself changed parts it would have been equally impossible. There are few men, indeed, who with such simple contrivances could have produced so amazing an effect on the minds of the sitters. Professional mediums would probably laugh at the clumsiness of his methods. But he had a power which they lack a power which more than compensated for any want of conjuring dexterity or experience in deceiving. He had such a control over the sitters' minds that he could divert their attention almost whenever he pleased; he could persuade them they had seen what they had not seen; he could make their very tests a trap into which they should themselves fall. And Mr. Wallace is quite right in supposing that this séance cannot be explained as a simple mechanical or sleight of hand trick. But the extra something which is wanting is not Spiritualism, it has no connection with Spiritualism. It is nothing else than the extraordinary genius of Mr. Davey.

Let us now turn to an account by Spiritualists of a sitting with Mr. W. S. Davis, to whom I have referred above (p. 295). Prior to his giving sittings himself, Mr. Davis had been very aggressive in denouncing bogus mediums, and, moreover, not a few persons were made aware that he had no "mediumistic" power at all. Further, the *New York Herald*, of June 13th, 1891, contained an article on the subject explaining what Mr. Davis proposed to do, and quoting some letters written by Mr. Davis to a reporter in proof of its assertions. Mr. Davis wrote, *inter alia*:

The great argument with the Spiritualists is this:—Are we deceived? Are we not as capable of detecting trickery as you are? We are shrewd in business matters, why should we be less shrewd in this?... Now, my object in giving these séances is to get evidence that people can be deceived very easily. ... All that I do is trickery, and I am doing just exactly what all of their famous mediums have been and are now doing.

This article and another warning article which appeared in *The Banner of Light*, a Spiritualistic paper, caused considerable disquiet among the Spiritualists who were endorsing Mr. Davis's manifestations as genuine. Finally, Mr. Davis "expressed a desire to give a séance under strictly test conditions and let a committee judge." This offer was accepted. In the following account I abridge from the statement made by Mr. Davis:—

The date for the test séance was July 23rd, 1891, and only 15 persons were permitted to attend. The séance was a success, and congratulations were plentiful. The following report appeared in *The Better Way* of August 15th, 1891:

AN ENDORSEMENT.
To the Editor of *The Better Way*.

DEAR SIR,—We, the undersigned, tested the powers of W. S. Davis on Thursday evening, July 23rd, 1891, by applying such conditions as, in our judgment, absolutely precluded the possibility of fraud. The medium submitted to severe tying. The only door leading into the séance-room was locked, sealed, and carefully watched. The medium permitted us to put him into a large bag so that not even his head was exposed, and the sealing, &c., was done by us, and not by confederates. We practically had charge of the séance from beginning to end, and there were no friends of the medium present not equally known to us. The room was carefully examined before the séance, and during the séance the cabinet was opened for examination at very frequent intervals. Under these stringent conditions wonderful manifestations of spirit-power were given without delay, and immediately after the cabinet curtain was drawn in each case.

It is worthy of note that writing was obtained on a slate which was locked up in a box and doubly secured by the liberal application of sealing wax, when the medium did not touch the box or slate.

James B. Bogert, Elizabeth F. Kurth, Mrs. M. T. Morris, Hermann Handrich, Wm. C. Coss, Elizabeth A. Smith, John M. Coombs, Eleanor Dailey, Elizabeth S. Davis, Alexander S. Davis, Louis Sherk, W. P. Munroe, Mrs. W. P. Munroe, Margaret Smith.

The following report also appeared in *The Progressive Thinker* for August 8th, 1891, and an account by Mr. Handrich was published in *Psychische Studien* for October, 1891:

"The undersigned, a correspondent for the *Psychische Studien*, of Leipsig, Germany, and a number of prominent Spiritualists and mediums of Brooklyn, received invitations from Mr. W. S. Davis to attend a séance under test conditions, in order to convince his friends of the genuineness of his mediumistic gifts. The handsome residence of Mrs. M. Towers, a lady highly esteemed by her friends, and of prominent social standing, had been placed at the disposition of Mr. Davis and members of the circle. As I have previously mentioned in a letter addressed to Mr. Davis (in relation to a former séance which was held in the same premises), that confederates, if any, could find an easy access behind the curtain which fenced off a corner of the large sitting-room, I was requested by the medium to seal the only door leading to the room from the vestibule of the house, to which I gladly consented. Double walls, trapdoors, and other hiding places are out of question; notwithstanding, I convinced myself by ocular inspection, and guided by Mr. Bogert, Inspector of Buildings. Next, I assisted in tying the wrists of the medium, and I know positively that it was not legerdemainly done, as the blood hardly could circulate under the firm pressure of a hard rope which was used for this purpose. A few minutes after the medium took his seat in the corner, hidden by the curtain which separated him from the spectators, brilliant sparks and lights appeared in front of it; no electric press-the-button affair, but genuine phenomena of occult power. Bells were ringing; one of them, and likewise a slate, were thrown over the curtain, which extended in height only about four feet from the floor. Mr. Wicks, the master of ceremonies, drew back the curtain. The medium extended his hands, which by close examination and bright light bore witness that they had not been untied, as the rope actually cut itself into the flesh, and the impressions of its texture were distinctly visible on the skin of the upper part of the wrists after the rope was cut through with a knife.

"Unbound, the medium again sat himself in the corner. The curtain was drawn; a long rope, besides the small one, was handed over the curtain by Mr. Wicks, and a few minutes after it was drawn back again and the medium was securely tied to the chair. Remaining in this position, his coat was taken off and put on again; musical instruments were played, accompanying the songs of the audience, and other manifestations witnessed, whilst the rope, after repeated examinations, proved not to have been tampered with in any way or manner,of which I gladly and conscientiously bear witness, as I convinced myself of the fact.

"The next test to which the medium submitted, after having been delivered of his fetters by the same occult intelligence which bound him to the chair, was to be put into a bag. I also assisted in this operation, and after the bag was shut and sealed over the medium's head, I knew that there was no other exit for him except by getting out where he was put in. In this position the medium was placed on his chair in the corner, and with him a small wooden chest containing an unprepared clean slate. The chest was thoroughly examined by me and others; the padlock, the cover, and buttons sealed up, and then placed in the corner where the medium was seated. After a short pause, Mr. Wicks and myself got the medium out of the bag and the slate out of the chest, and found written thereon: 'We have done enough. H.' (H. stands for Haicidoka, the *nom de plume* of the medium's control.) The seals on the door, padlock, bag, and chest were found unbroken, to which I testify without fear or favour."

HERMANN HANDRICH.

I will now describe just exactly what did occur at the test séance.

After the company was seated, I took a seat in the "cabinet." Mr. Wicks asked the people to sing. While they were singing "Nearer, my God, to Thee," Mr. Wicks extinguished the light and handed to me, in the darkness, an electric battery, to which was connected by a wire a small incandescent lamp. Lights were shown by "pressing the button," although Mr. Handrich states that such was not the case. Various results were produced by moving the lamp from one place to another, by short and long pressure on the button, and by wrapping different coloured tissue paper around the lamp, &c.

Mr. Wicks then gave me a "transparency" which had been hidden behind a large picture hanging on the wall. The lamp was placed inside of the thin box. The button was pressed and a life-sized hand was shown. This was the end of that part of the show, and in a few minutes Mr. Wicks asked if a little singing would increase the forces, and was answered by two raps, meaning "don't know." Being anxious to get more phenomena they sang, which enabled Mr. W. to get the lamp and "transparency" away from me without being heard.

After the room was lighted, the cabinet curtain was removed and I came out of the pretended trance and stated to the company that conditions seemed to be favourable and that we would commence to test matters. …

Mr. Handrich and Mr. Bogert fastened the door shut by locking it and sticking strips of paper from the door to the door-case with sealing wax. ...

The "first test" consisted of being tied by occult agency, having my coat removed without tampering with the ropes, and permitting the company to examine the fastenings as long and as often as desired.

After I got well "under control" Mr. Wicks hid me from view by drawing the curtain. He then threw over the curtain a long and a short rope. The short rope was not used, but was substituted by another rope carefully knotted, &c. This prepared rope is well arranged, it generally requiring considerable time to make it. Braided cotton sash cord is used. The knots are made when the rope is wet and are held together by driving in soft smooth iron nails which are filed off close and hidden by working the rope over them. This is the "Spirit Tie Harness." I got it from Frank Vanderbilt, an old time medium who flourished when the Eddy family, Eva Fay, the Davenport Brothers, and many others were in the business. ...

In my tie the principal knot is in plain view and people think it is the last knottied and they know it is a slip knot. They never for a minute seem to think that any other knot is of any importance. After a very careful examination of the ropes it is generally settled that the trick is in my coat, if the investigator happens to be of the opinion that it is a trick.

In talking with me about this particular manifestation, the Spiritualists argue int his way: "Kellar and others do this in the theatres, but they do not believe that Kellar could do it under test conditions unless he is a medium—and where is the proof that he is not a medium?"

The next "test" consisted in permitting the company to bind my wrists together with stout twine. When I announced that I would submit to this test, Mr. Handrich took from his pocket a thin copper wire covered with cloth, and asked permission to bind me with it. I said certainly, and walked over to him remarking that it was immaterial to me who did the tying or brought the string. He said it isn't string, it is wire. I said I thought that metals of any kind would produce very uncomfortable sensations in my wrists, as wire would act as a conductor of electricity. He apologised and said that the twine would serve the purpose fully as well. ... I then bared my left arm and asked Mr. Bogert to tie the twine around the wrist. He did the tying so loose that I could have pulled my hand through. I told him to untie it and make a better job of it. He declined to do so, saying that he had implicit confidence in me and would not stop the circulation of blood for anybody. (You see he would have been as good as a confederate, so far as the tying is concerned, and it would not have been necessary for me to resort to the regular trick of stealing slack either.) I pretended to show a little temper, and said it was a test séance, and that I wanted everything done thoroughly. Mr. Bogert then tied the cord around my wrist very tightly,—the knot being on the inside of the wrist. I presented my left arm with the palm of the hand up. I then went to another person and had another knot added to the ones Mr. Bogert had made—I went to other persons, allowing each to make more knots. All of this looked as though I was being well tied, but my real object in going from one to another to have more knots made was to take a turn around my wrist with one end of the twine. This is sleight of hand, and is not noticed when done properly. One end of the twine is wound around the wrist after the last knot on the left wrist has been made, just as I am in the act of telling a person to take the other end and pull on it; then after taking the hitch on the first end I ask another person to pull on that end. Then I put my right arm exactly over my left and request the two persons to bring the two ends together and tie them tightly. Then I go from one to another, asking each person to add a knot. Then the knots are fastened with sealing wax. I then take a seat in the cabinet and sit there in full view until I get "under control," then the curtain is drawn. I immediately turn my arms in opposite directions, which takes the turn out of the rope, and my hands are free. I can take my hands out of and get back into the fastenings with great quickness, which I did on this occasion.

The next test consisted of being tied up in a bag. The end of the bag is gathered together by passing a twine through brass eyelet-holes and drawing the bag together by pulling the two ends of the string. When the bag is being drawn up over my head, I catch one of the loops of the twine and pull in as much slack as I can. This is similar to Mrs. Martin's neck-tie exposed by you [At a rope-tying materialisation séance which I attended in New York.—R.H.], but is much better, since I do not have to close a door to pull in slack. Mr. Wicks is prepared to erase any marks put on the ropes, and after the fastenings are pronounced intact he immediately cuts the rope in a number of places while he holds the bag up and while I am getting the extra rope in my pocket. The bag tie knot is also fastened with wax. This looks good to the audience and it apparently increases the value of the test, but the sealing is really a help to the trick, as it makes it impossible to untie the knot after the performance, and the string has to be cut.

The next "test" was to get writing on a slate locked up in a box while I was in the bag. You should have seen them examine the slate. They washed it very carefully; held it against a light to burn off a possible prepared message written in sympathetic inks. Little or no attention was paid to the box. The slate was put into the box. The box was then closed and locked. The key-hole of the lock was covered by a piece of paper which was fastened with sealing wax. Several strips of paper were wound around the padlock and fastened with sealing wax. Then the lid and box were fastened together by sticking paper, &c.

After I got "under control" again (in the bag) and while they were singing another hymn, the lights were lowered, the cabinet curtain drawn, and the box was put into the cabinet. All that I had to do was to get my head and arms out of the bag, and push in one end of the box and take the slate out. [Here Mr. Davis describes the trick-box, with diagrams.—R. H.]

The most singular thing in connection with this "test séance" is the fact that the tests were all "forced" and nobody thought of it. But why should they have thought of it since nearly everything called "tests," so far as public mediums are concerned, are forced? The manifestations occurring while I was tied in the bag, &c., consisted of performing on the banjo, flute, violin, bones, writing on slates, exhibiting tambourine, bells and hands over the curtain, &c.

Mr. Davis, at my request, repeated the above performances in detail before a group of our members, including myself, in New York, and after each performance he illustrated his methods in full light to the complete satisfaction of all the persons present. Mr. Davis informs me that he and Mr. Wicks "will undertake to explain the methods of, and reproduce the performance of, any slate-writing, fire-test, rope tying, etherealising, and materialising medium who will go before a committee of your Society and permit us to be present."

Returning now to the accounts which I have given of Mr. Davey's methods,—in connection with the statement made by Mr. Wallace that unless *all* Mr. Davey's phenomena can be explained by "trick," he will be confirmed in his "belief that Mr. Davey was really a medium as well as a conjurer," what is the value of the testimony to such phenomena given by persons who are in the position of Mr. Wallace, unable to distinguish between the avowed medium and the avowed and proved conjurer? Many other Spiritualists have apparently held the same position. Thus, Mr. Dixon, writing to *The Spiritualist* in 1875, said that he was "thoroughly nonplussed" by Dr. Lynn's cabinet performance.

Unless the spirits did "it," I am utterly at a loss how to account for it, and my only way out of the difficulty, when questioned by my friends, was to claim this part of the performance as a genuine piece of Spiritualism. ... I am sure the cause of Spiritualism would gain immensely if Lynn's séance [?] could be explained. ... If I had not paid my money for an evening with Dr. Lynn I should have come away from it as a Spiritualist séance with the most perfect assurance that the manifestations were genuine.

Mr. Gledstanes, writing in the same year, 1875, points to the probability of the medium's *doubles* doing the feats, and strongly suggests that Maskelyne and Cook project arms from their bodies! Mr. Coleman, writing in 1874, says:—

All inquirers who desire to study the psychological character of spiritual manifestations should be recommended to visit Messrs. M. and C. [Maskelyne and Cook], who have gone on practising them with a perseverance worthy of a better aim, and who are, in my opinion, the best of living mediums.

Somewhat similarly, Mr. Wallace in 1877, describes a performance of Dr. Lynn's at the Royal Aquarium, including a cabinet trick, and the moving and floating of a table about the stage, two feet from the floor. He says:—

Mr. Davey's Imitations by Conjuring, &c.

Your readers must be told that Dr. Lynn is not the performer; but a gentleman who is introduced as "a medium—a real medium"; and I must say I believe him to be one. ...

A week later "M.A. (Oxon.)" added his testimony as follows, in a passage which I quoted in my previous article:—

I am glad to see that Mr. Alfred Wallace agrees, after seeing Lynn's medium, with the substance of my letter in your issue of July 6th. Given mediumship and shamelessness enough so to prostitute it, and conjuring can, no doubt, be made sufficiently bewildering. It is sheer nonsense to treat such performances as Maskelyne's, Lynn's, and some that have been shown at the Crystal Palace, as "common conjuring." Mr. Wallace positively says, "If you think it is all juggling, point out exactly where the difference lies between it and mediumistic phenomena."

It is not surprising that statements of this kind should have called forth such an emphatic remonstrance as the following from Mr. Coates, who at the same time professed his belief in Spiritualism:—

The man who cannot distinguish between mediumship and conjuring, though he be a doctor of law, science, medicine, or divinity — his evidence is shaky, his theories not worth the paper they are written on, and his advocacy the cause had better be without.

Mr. Wallace, however, does not seem to have made any advance since 1877 in his discrimination between conjuring and alleged medium ship of the kind we are considering. Why? Is it not obviously because no discrimination can be made? And accordingly many Spiritualists continue to call conjurers "mediums." A recent instance occurs in a communication to *Light* for October 24th, 1891, by "T. W." He writes that "we may say, without prevarication, that the conjurers have utilised physical Spiritism." He refers to a coin trick by Bosco, and apparently regards it as involving "mediumship." (It was probably a variation of the coin trick described in *Modern Magic*, p. 161.) He also mentions the "famous conjurer," M. Duprez, and says: "He must be a powerful physical medium. I saw his performance some years ago, and ... I believe that there is scarcely a 'trick' performed in which he is not aided or supplanted by unseen force." Now, statements like these, absurd as they seem to persons familiar with conjuring operations, do not originate simply and merely from ignorance of conjuring, and the remedy is not, simply and merely, to become acquainted with certain trick-devices.[4] The remedy for such absurdity is to learn that an un-

---

[4] Of course a knowledge of trick-devices is likely to make a witness hesitate in many cases before concluding that a certain "phenomenon" was not produced by ordinary means. But it sometimes acts the other way, especially if the witness thinks that he is already an expert. The well-known conjurer Kellar was not familiar with special "slate-writing" methods when he was in India in 1882, and Eglinton was able to deceive him. Kellar afterwards changed his opinion as to Eglinton's phenomena when he became familiar with the methods. Yet even at this later date, he was unable, in conversation with me, to offer any explanation of the production of writing between slates screwed together and sealed, &c., as described above, though he did not suppose it was other than a trick. Similarly it seems to me quite likely that an expert in the different trick-methods of opening sealed envelopes might be baffled by the trick described as follows in *Revelations of a Spirit-Medium*, pp. 178-9:

"But the smoothest thing in the sealed letter reading, and the one that has puzzled the people for years, is usually done in connection with 'slate-writing.' The 'sitter' is furnished with a heavy white envelope, of small size, and a white card of the size of an ordinary visiting card. He is requested to write the name of a spirit friend on the card, and to write one or not more than two questions with it. After he has written as requested, he is instructed to place the card in the envelope with the writing next the smooth side and away from the glue. This being done, he is furnished with letter-wax with which he seals the seams to prevent the envelope being opened.

"The 'medium' now takes his seat at the table opposite his sitter and near a window. Placing the envelope on a slate he thrusts it beneath the table. After sitting long enough to do his work, raps are heard on the slate, and, withdrawing it, he hands it to the 'sitter.' The envelope still lies on the slate, and there is no evidence of its having been touched. The seals are intact, and there is not a mark or mar on it.

"On the slate is written the replies to his questions, and the name of the spirit addressed is signed at the bottom of the message."

The expert in opening letters would rightly conclude that the envelope had not been opened, and if he did not allow a margin for his ignorance of special devices, he might be disposed to attribute the phenomenon to some "clairvoyant" power.

"In order to perform this trick, do just as the 'medium' did up to the time he placed or held the slate beneath the table. Instead of holding it there with your hand, slip one corner between your leg and the seat of the chair. Thus you are holding it by sitting on it. Your hand is now free to do as you choose with. Your 'sitter' cannot see your movements, for the table is interposed. Put your fingers into

initiated witness cannot describe to himself the real conditions under which the feats were performed, because his powers of observation and memory are inadequate for the purpose. He may *rightly* conclude that under such and such conditions the feats are inexplicable by conjuring, but he *wrongly* concludes that the conditions were as he describes them.

The plain result from our investigation is that the great bulk of the testimony to the "physical" marvels of Modern Spiritualism is not entitled to serious consideration as affording any evidence of super normal phenomena. I may conclude with a warning which I venture to give specially to our members in America, viz., that nearly all professional mediums form a gang of vulgar tricksters who are more or less in league with one another. Associated with this combination, here and there, are certain other persons who either have been, or intend to be, professional mediums, and who are equally untrustworthy. These tricksters are continually deceiving fresh groups of uninitiated observers of their performances, and I frequently receive accounts of them which, I need hardly say, are entirely worthless for the purposes of our investigation. It is not from the professional mediums—so numerous in the United States—for "slate-writing," "materialisation," and kindred performances, that we can look for any enlightenment whatever, on the positive side, in the course of Psychical Research.

---

the ticket pocket of your coat and bring out a small sponge that is saturated with alcohol; dampen the envelope over the card and you can easily read the name and question. Write the answer and sign the name addressed, and your 'sitter' will be 'paralysed' with astonishment.

"Nothing will serve to dampen the envelope but alcohol. Nothing else will allow of your reading the writing on the enclosed card, and nothing else will dry out quickly enough and leave absolutely no traces of any manipulation. Water will not dry out quick enough, and when it does dry leaves the envelope shrunken where it was applied, thus leading your 'sitter' to suspect that you have not played fair."

Some Suggestions for Further Reading and Additional Resources

For researchers and interested readers who want to explore more ideas about magic, science, and fraudulent mystics, here's a very non-exhaustive annotated list of additional books and resources compiled by Rev. Dr. Matthew L. Tompkins.

Books on Fraudulent Mediumship Methods

Abbot, D.P. (1912). *Behind the Scenes with Mediums*. The Open Court Publishing Co.
> Written by magician David P. Abbot includes detailed explanations of fraudulent mediumistic tricks, including texts from exposes that Abbot originally published in *The Journal of the American Society for Psychical Research*, a group that Hodgson was instrumental in running in the years after he published the Mal-Observation Report.

Keene, L. & Spraggett A. (1976) *The Psychic Mafia*. St. Martin's Press.
> An interesting update to the other Victorian-era exposés in this list. This particular work follows the same pattern of an ex-medium revealing fraudulent methods, except Lamar Keene was performing his con in the mid 20th century. More recently in 2022, Keene was the subject of a BBC Radio 4 series from investigative journalist Vicky Baker, titled *Fake Psychic*, that is also well worth a listen.

Home, D. D. (1878). *Lights and Shadows of Spiritualism*. Virtue and Co.
> Unlike the other items on this list, this book was written by a man who was a very active and self-proclaimed genuine medium, but nonetheless details a variety of fraudulent methods. For helpful context, this pairs very nicely with Peter Lamont's (2005) biography of Home.

Medium, A. (1882). *Confessions of a Medium*. Griffith & Farran.
> A pseudonymously written expose of fraudulent medium techniques and methods. Recommended by Richard Hodgson.

Oxon, M. A. (W. S. Moses). (1882). *Psychography: A treatise on one of the objective forms of psychic or spiritual phenomena*. Psychological Press Association.
>A book on the phenomena of independent slate writing, written by William Stainton Moses, a believer in the genuineness of slate writing phenomena.

Price, H. & Dingwall, E. J. (1922). *Revelations of a Spirit Medium*. E.P. Dutton and Co.
>A reproduction of the anonymously authored Revelations of a Spirit Medium that was originally published in 1891. This book includes an additional preface, notes, and an extensive bibliography that were added by psychical researchers Harry Price and Eric Dingwall.

Robinson, W. E. (1898). *Spirit Slate Writing and Kindred Phenomena*. Munn & Co.: New York.
>Written by William Robinson (aka Chung Ling Soo), this book contains numerous methods for slate writing and other fraudulent mediumistic tricks. Original copies are highly sought by collectors and very expensive, but Curious Publications offers an excellent facsimile edition.

Truesdell, J. W. (1884). *The Bottom Facts Concerning the Science of Spiritualism: Derived from careful investigations covering a period of 25 years*. G. W. Carleton & Company.
>An exposé of fraudulent medium techniques and methods written by a former medium, includes detailed explanations of hoax spiritualistic performances. Recommended by Richard Hodgson.

Wiseman, R. J., & Morris, R. L. (1995). *Guidelines for Testing Psychic Claimants*. University of Hertfordshire Press.
>A practical treatise aimed at parapsychological researchers wishing to conduct tests of alleged psychics, co-written by magician and psychologist Richard Wiseman and parapsychologist Robert Lyle Morris.

And Additional Resources

Books on the History Psychical Research

Blum, D. (2007). *Ghost Hunters: William James and the search for scientific proof of life after death.* Penguin.
   A narrative non-fiction book about the psychical research of the legendary American psychologist William James. It includes some excellent details about his relationship with Richard Hodgson.

Baird, A. T., (1949). *Richard Hodgson: The story of a psychical researcher and his times.* Psychic Press.
   The most extensive existing biographical account of Richard Hodgson's life, which sometimes borders on hagiography. The author is a firm believer in the existence of life after death, which sets a distinct tone (the book continues after Hodgson's death with accounts of his spirit's activities).

Hartzman, M. (2021). *Chasing Ghosts: A tour of our fascination with spirits and the supernatural.* Quirk Books.
   An illustrated narrative-nonfiction tour of society's relationship with ghosts ranging from antiquity to the present day, featuring a variety of fascinating case studies presented with a great mix of primary sources, scholarship, and pop cultural references.

Lamont, P. (2005). *The First Psychic: The peculiar mystery of a notorious Victorian wizard.* Little, Brown.
   A thoroughly researched and referenced biography of the notorious celebrity medium Daniel Dunglas Home.

Lamont, P. (2013). *Extraordinary Beliefs: A historical approach to a psychological problem.* Cambridge University Press.
   A contemporary academic account of how Modern Spiritualism influenced developments in the field of psychology.

Larson, E. (2007). *Thunderstruck.* Crown.
   A narrative non-fiction book about development of wireless telegraphy, including some very interesting stories about how these developments were interwoven with psychical research.

Luckhurst, R. (2002). *The Invention of Telepathy, 1870-1901*. Oxford University Press.
    A historical exploration of the concept of mind-to-mind communication with some excellent historical details and context about fraudulent mediumship and scientific testing.

Noakes, R. (2019). *Physics and Psychics: The occult and the sciences in modern Britain*. Cambridge University Press.
    A contemporary academic account of how Modern Spiritualism influenced developments in the field of physics.

Sera-Shriar, E. (2022). *Psychic Investigators: Anthropology, modern spiritualism, and credible witnessing in the late Victorian age*. University of Pittsburgh Press.
    A contemporary academic account of how Modern Spiritualism influenced developments in the field of anthropology.

Wiley, B. H. (2012). *The Thought Reader Craze: Victorian science at the enchanted boundary*. McFarland.
    Written by a historian and magician, this book explores the relationship between spiritualists, thought readers, and scientific investigators, with a particular emphasis on the relationships between entertainers and researchers.

Relevant Books on Cognitive and Anomalistic Psychology

Chabris, C., & Simons, D. (2010). *The Invisible Gorilla: How our intuitions deceive us*. Crown.
    A popular science book written by the originators of the Invisible Gorilla Experiment. Details the real-life implications of metacognitive illusions and inattentional blindness.

Mack, A., & Rock, I. (1998). *Inattentional Blindness*. MIT Press.
    Details the flagship experiments that established inattentional blindness as a psychological phenomena.

French, C. C., & Stone, A. (2017). *Anomalistic Psychology: Exploring paranormal belief and experience.* Bloomsbury Publishing.
>An excellent academic textbook on the field of anomalistic psychology, providing a great overview of the field along with excellent references.

French, C. (2024). *The Science of Weird Shit: Why our minds conjure the paranormal.* MIT Press.
>A popular psychology text on anomalistic psychology. A great accessible companion piece to French & Stone's (2017) textbook.

Simons, D., & Chabris, C. (2023). *Nobody's Fool: Why we get taken in and what we can do about it.* Basic Books.
>A follow-up to Chabris & Simon's (2010) *Invisible Gorilla* book, this piece broadens the focus to the concept of scams and hoaxes (including psychic fraud), and also offers practical scientifically-grounded suggestions on how to guard against being deceived.

Wiseman, R. (2011). *Paranormality: The science of the supernatural.* Macmillan.
>An excellent popular science account of anomalistic and parapsychology, written by Richard Wiseman, who is both an academic psychologist and a magician

Books on The Science of Magic

Lamont, P., & Wiseman, R. (1999). *Magic in Theory: An introduction to the theoretical and psychological elements of conjuring.* Univ. of Hertfordshire Press.
>Written by two magicians who are also (respectively) a psychologist and a historian. This book is designed to distill concepts from the vast field of performance magic literature to help parapsychologists better understand the risks of fraudulent psychic claimants. This book arguably laid the foundation for the contemporary 'science of magic renaissance,' which has been marked by an unprecedented number of academic publications on the topic of performance magic.

Further Reading

Kuhn, G. & Pailhes, A. (2023). *Psychology of Magic: From lab to stage.* Vanishing Inc.
> A book on the science of magic written for magicians. Gustav Kuhn and Alice Pailhès have written a guidebook for performers on how to integrate empirical research findings into their own trick designs and presentations.

Kuhn, G. (2019). *Experiencing the Impossible: The science of magic.* Mit Press.
> A great accessible text about how the scientific study of performance magic can provide insights into psychological mechanisms. I think this is currently one of the best introductions and overviews of contemporary science of magic research.

Lutrell, A. (2015) *Psychology for the Mentalist.* Mind Tapped Productions.
> A guide for performers on how to integrate empirical psychological principles to enhance their own presentations of theatrical mind reading.

Tompkins, M. L. (2019). *The Spectacle of Illusion: Magic, the paranormal & the complicity of the mind.* Thames & Hudson.
> An illustrated narrative non-fiction book about historical and contemporary relationships between magicians, fraudulent mystics, and scientific researchers. A selectively honest account about lies about the truth about lies, with lots of weird pictures.

Additional Resources: Libraries, Archives, and Relevant Organizations

The Internet Archive
archive.org/
   Many of the older works listed above are out of copyright and digital versions are freely available online via the Internet Archive's excellent online library.

The International Association for the Preservation of Spiritualist and Occult Periodicals (IASOP)
iapsop.com
   A fantastic digital preservation project that provides free access to text searchable scans of Spiritualist and occult periodicals published between 1815 and 1939.

The Society for Psychical Research (SPR) Libraries and Archives
1 Vernon Mews, London, W14 0RL, United Kingdom
spr.ac.uk/about/libraries-and-archives
   The SPR of Hodgson and Davey's time has been continuously active until the present day. Their physical collection is housed in London, and their digital library includes all the SPR's *Journals* and *Proceedings* from 1882-present. Full access to their excellently maintained archive is available to members, but non-members can search the catalogues and abstracts and they can also access the full digital library through a free trial.

The Society for Psychical Research Collection at Cambridge University Library
Cambridge University Library, West Road, Cambridge CB3 9DR, UK
lib.cam.ac.uk/collections/departments/archives-modern-and-medieval-manuscripts-and-university-archives-0
   In addition to the Vernon Mews library of the contemporary SPR, many of the society's historic materials are held at the University of Cambridge. This collection includes correspondence between members as well as detailed records of their investigations.

Further Reading

The Harry Price Library of Magical Literature
Senate House Library, University of London, Malet Street, London, WC1E 7HU, United Kingdom
london.ac.uk/about/services/senate-house-library/collections/printed-special-collections/harry-price-library-magical-literature
    This is one of my personal favorite library collections. Held at the University of London's Senate House Library, The Harry Price Library encompasses nearly 13,000 items related to magic, psychical research, and psychology including books, pamphlets, periodicals, and ephemera, dating from 1472 up to through the 21st century. Come for the periodicals, but stay for the hair samples from the allegedly talking mongoose.

Conjuring Arts Research Center
11 W 30th St, New York, NY 10001, United States
conjuringarts.org/library/
    A non-profit library and research center devoted to performance magic and its allied arts, with a collection that spans over 12,000 volumes. The Center also maintains an extensive digital collection that can be accessed at https://askalexander.org/

The Magic Circle Library
Centre for the Magic Arts, 12 Stephenson Way, London NW1 2HD, United Kingdom.
themagiccircle.co.uk/object/the-magic-circle-library/
    The Magic Circle Club and Magic Circle Foundation are, respectively, private a club for professional magicians and a non-profit charity devoted to educating the public in the history and practice of the art of performance magic. Non-members can make research appointments at the library to view its collection of 12,000 manuscripts, books, and magazines.

Harry Houdini Collection at the Library of Congress
101 Independence Ave SE, Washington, DC 20540, United States
    Contains over 4,000 items related to Houdini, many bequeathed from his own personal collection including correspondence, scrapbooks, paraphernalia, and books.
    Many digitized items are available through the HathiTrust (filter

And Additional Resources

your search for "Harry Houdini Collection (Library of Congress)"
https://catalog.hathitrust.org/

The Science of Magic Association SoMA
https://scienceofmagicassoc.org/
SoMA is an interdisciplinary organization that promotes rigorous research directed toward understanding the nature, function, and underlying mechanisms of magic. The idea is promote collaborations between performers and academics. We also maintain a semi-regular newsletter, and hold both virtual and live events around the world.

www.ingramcontent.com/pod-product-compliance
Lightning Source LLC
Chambersburg PA
CBHW020451030426
42337CB00014B/1506